Urban Sustainability

The Urban Sustainability Book Series is a valuable resource for sustainability and urban-related education and research. It offers an inter-disciplinary platform covering all four areas of practice, policy, education, research, and their nexus. The publications in this series are related to critical areas of sustainability, urban studies, planning, and urban geography.

This book series aims to put together cutting-edge research findings linked to the overarching field of urban sustainability. The scope and nature of the topic are broad and interdisciplinary and bring together various associated disciplines from sustainable development, environmental sciences, urbanism, etc. With many advanced research findings in the field, there is a need to put together various discussions and contributions on specific sustainability fields, covering a good range of topics on sustainable development, sustainable urbanism, and urban sustainability. Despite the broad range of issues, we note the importance of practical and policy-oriented directions, extending the literature and directions and pathways towards achieving urban sustainability.

The series will appeal to urbanists, geographers, planners, engineers, architects, governmental authorities, policymakers, researchers of all levels, and to all of those interested in a wide-ranging overview of urban sustainability and its associated fields. The series includes monographs and edited volumes, covering a range of topics under the urban sustainability topic, which can also be used for teaching materials.

Biman Gati Gupta
Editor

Urban Development with Treated Recycled and Reuse of Wastewater: Sustainable Management

Springer

Editor
Biman Gati Gupta
Department of Civil Engineering
Elitte College of Engineering
Maulana Abdul Kalam Azad University
of Technology
Sodepur, West Bengal, India

ISSN 2731-6483 ISSN 2731-6491 (electronic)
Urban Sustainability
ISBN 978-981-96-0856-0 ISBN 978-981-96-0857-7 (eBook)
https://doi.org/10.1007/978-981-96-0857-7

This Springer imprint is published by the registered company Springer Nature Singapore Pte Ltd.
The registered company address is: 152 Beach Road, #21-01/04 Gateway East, Singapore 189721, Singapore

Contents

About the Editor

Dr. Biman Gati Gupta, Ph.D., C.Engg., BCEEM, FIE, FEWRI, FASCE passed B.E. (Civil Engineering) from B. E. College, Calcutta University, India, in 1974 and then Completed Ph.D. (Ecological Studies) from Kalyani University, Kalyani, India, in 2017. Completed short course on Globalization and Public Health Governance from Imperial College, London, UK.

Biman Gati Gupta has more than 47 years of experience in the academy (Professor and HOD), research, consultancy (Vice-President of a Consultancy House in India), and industrial activities on environmental engineering and civil engineering projects. Biman Gati Gupta is listed in the Who's Who 2021 published by AAEES, USA. Data Base: Biman Gati Gupta worked on environmental science, environmental engineering, air pollution and its effect on health, soil degradation, effects on fruits and vegetables, effects on public health, and socioeconomic activities in the industrial cluster. Biman Gati Gupta's papers published in Springer, Elsevier, Sage, and other journals.

Biman Gati Gupta associated as Project Management Expert on Civil Construction of Power Projects, preparation of Detailed Project Reports on Infrastructure, urban planning, food parks, IT park, SEZs, EPIPs, commercial complexes, textile parks, jute parks, automobile parks, foundry parks, etc. including meeting environmental regulations.

Biman Gati Gupta is Remediation, EIA, and Eco-Planning of various large, medium projects. Financial analysis and feasibility studies of different projects, negotiation with the client, Govt. Department and World Bank for approval.

Extensively worked on Public health deterioration due to the use of contaminated water, taking metal-induced fruits and vegetables in the urban area, and inhaling smog in textile clusters.

Presently, Biman Gati Gupta is a council member and trustee of Textile Institute, Manchester, England, Fellow of the American Society of Civil Engineers, USA, Fellow of Environment and Water Research Institute, USA, Fellow of Institute of Engineers, India, and Eminent Member of Environmental Engineers and Sciences, USA.

Urban Planning and Design

Biman Gati Gupta

The political and technological process of urban planning focuses on the layout and use of land inside and around cities, as well as the air, water, and infrastructure that flows into and out of them, including distribution and transportation networks. Some of the oldest known urban sites show signs of planning, hence the history of urban planning is closely related to the history of the city. Many cities in the pre-Classical and Classical eras had predetermined layouts, but most grew naturally. See also: Urban planning in ancient Egypt for more on the built environments that were typical of the Harrapan, Minoan, Mesopotamian, and Egyptian civilizations that existed in the third millennium BC. The first recorded description of urban planning appears in the Epic of Gilgamesh: "Go up on to the wall of Uruk and walk around. Inspect the foundation platform and scrutinise the brickwork. Testify that its bricks are baked bricks, And that the Seven Counsellors must have laid its foundations. One square mile is city, one square mile is orchards, one square mile is claypits, as well as the open ground of Ishtar's temple. Three square miles and the open ground comprise Uruk. Look for the copper tablet-box, Undo its bronze lock, Open the door to its secret, Lift out the lapis lazuli tablet and read" [1].

1 Ancient History

Based on their unique urban planning features, archaeologists consider the Indus Valley civilization sites of Mohenjo-daro, Lotha, Dholavira, and Harappa to be the first known instances of planned and managed urban areas. These sites are located in what is now northwest Pakistan and India. [2, 3] Many of these early towns'

B. G. Gupta (✉)
Elitte College of Engineering, MAKA University of Technology Sodepur, Kolkata 700113, India
e-mail: bimangupta@yahoo.in

B. G. Gupta (ed.), *Urban Development with Treated Recycled and Reuse of Wastewater: Sustainable Management*, Urban Sustainability,
https://doi.org/10.1007/978-981-96-0857-7_1

streets were paved and arranged in a grid form, with main thoroughfares leading down to smaller residential lanes. The layout of many Harrapan dwellings was designed to provide domestic seclusion and shield inhabitants from outside disturbance, according to archaeological evidence. Additionally, many of these residences had their own water wells, which were likely used for both ceremonial and hygienic functions. Drainage systems, apparently associated with a highly developed concept of urban cleanliness, were common in these ancient towns, making them distinctive [2]. The development of grid-based urban planning may have evolved from rectangular-field agriculture [4].

During the late Post classic era, the core parts of most Mesoamerican towns were well-organized and usually had public plazas surrounded by public structures. Residential areas in the vicinity, on the other hand, usually exhibited little to no evidence of planning [5].

2 Development of East Asia

Urban planning in China has a long history that spans millennia.

Growth of ancient China

A form of town called Jōkamachi originated during the feudal era in Japan, however certain towns like Nara and Heian-kyo adhered to ancient Chinese design principles. Built for and focused on defence, the settlements were known as castles. Castles and other structures were often positioned such that they could be concealed by the closely packed houses around them. Roads were constructed to lengthen the routes leading to the castles. Castle towns like Edo (which became Tokyo) are common.

3 Greco-Roman Culture

The orthogonal urban arrangement is often believed to have been initially proposed by the Greek philosopher Hippodamus (498–408 BC).He was actually considered a pioneer in urban planning, an honour that Aristotle bestowed upon him, and it remained in use well into the twentieth century. This is, however, only partly justified as Greek cities with orthogonal plans were built long before Hippodamus [8, 9]. Hippodamian planning, after which he was named, is an orthogonal arrangement of roughly square street blocks. The fact that Hippodamus was not the actual creator of this arrangement is shown by several archaeological discoveries, including those from ancient Egypt [10]. Perhaps the first instance of criticism of urban planning is Aristotle's (Politics 2.8) criticism and mockery of Hippodamus.

Newly founded towns and cities with more or less regular orthogonal layouts were the centres of Greek colonisation throughout the Mediterranean shores beginning in the late eighth century. Over time, the new designs started to seem more consistent [11]. The regular shape of the city of Miletus was supposedly dictated by the thoughts of Hippodamus of Miletus after its destruction by the Persians in 494 BC [12]. It seems that new colonial settlements and those reconstructed quickly after disaster were more prone to regular orthogonal layouts.

A century after Hippodamus, Alexander hired the architect Dinocrates to design his new city of Alexandria, the pinnacle of idealised ancient Hellenistic urban planning, with a level site near the Nile's mouth facilitating the city's regularity.

The ancient Romans built their colonies on top of regular orthogonal constructions as well [13]. It is likely that the regularly designed towns of the Etruscans in Italy and other Hellenic and Greek models served as inspiration for them [14]. Good design principles were first out by the Roman engineer Vitruvius, and they are being used today [15].

In order to make city planning easier for the general populace, the Romans utilised a unified system. A rigid, rectilinear grid of streets around the centre forum, which housed the city's services. Water, transportation, and sewage disposal were all made possible by rivers that ran beside or through the city at one point [16]. Across their vast empire, the Romans constructed hundreds of towns and cities. The ruins of these plans may be seen in many European places, including Turin, and they demonstrate the Romans' very rational approach to city planning. They planned to create a square grid with streets laid out at right angles to each other. The width and length of every road was identical, with the exception of two that were somewhat wider. In the heart of the grid, the east-west decumanus and the north-south cardo crossed. Flagstones were meticulously set and then filled in with smaller, tightly packed boulders and pebbles to form all of the roadways. Where necessary, bridges were built. Insulas were the Roman version of current city blocks, and they were squares defined by four roadways. About 80 yards (73 metres) squared made up each island. Over time, the city's layout might become more complex, with streets and lanes winding their way between buildings of varying sizes.

A wall may have formerly encircled the city, delineating its boundaries and providing protection from outsiders. Farmland was left undeveloped in areas outside of the municipal boundaries. Huge gateways topped with watchtowers marked the termini of each major route. While besieged, the city had a portcullis built above the entrance and extra watchtowers built around the city walls. The city fortifications were bypassed by an aqueduct.

Since the Romans and Greeks are often considered the primary forebears of contemporary Western civilization, the topic has received a lot of attention, and the development of Greek and Roman urbanisation is very well-known due to the abundance of recorded materials. It must not be overlooked, however, that other European civilizations, mostly of Celtic descent, also had urban centres [17]. There are some instances that seem to have been premeditated, such the Polish town of Biskupin in the Lusatian zone.

4 Primitive Europe (500–1400)

The cadastral plan from 1830 served as the basis for this plan of Elburg in the Netherlands. In 1392, Arent toe Boecop, who was a steward of the duke of Gelre, built Elburg. It seems that Arent was involved in private enterprise. He had acquired a plot of land next to the current town, and with his lord's blessing, he planned to rebuild and expand it, as well as relocate the local inhabitants by selling them home lots. A canalised river and an intersecting roadway form the focal point of the very symmetrical design. On the other hand, the symmetry is broken by the church in the eastern corner and the pre-existing street on the northwest side—the only curving street in the whole town. The broad outer ditch and corner bastions were constructed in the latter part of the sixteenth century.

Primitive Roman culture

Little evidence of urban civilization was left in western and central Europe as the West Roman Empire began to crumble in the 5th century and was further devastated in the next 500 years by invasions by the Huns, Germanic peoples, Byzantines, Moors, Magyars, and Normans. But it seems like things got better economically and politically in the tenth and eleventh century. As a result, the city's monetary economy and cultural life were able to recover, and commerce and craft flourished. In the beginning, urban culture flourished in pre-existing communities, sometimes amid the ruins of ancient Roman towns and cities. However, as time went on, more and more towns sprung up. At the same time, the cultivated land area expanded in tandem with the fast population growth in Western Europe. New towns and villages were established in previously uncultivated lands as the foundation for reclamations, while agricultural sections of already established communities were expanded [18].

During the early Middle Ages, urban expansion took place "like the circular rings of a tree" [19] in both the heart of bigger cities and enlarged villages, with a concentrate on fortified abbeys, fortresses, and (sometimes) abandoned Roman nuclei. The city's layout became more organic, mimicking the forms seen in agricultural terracing, to accommodate the new center's frequent placement on elevated, defendable territory.

(Wales) Caernarvon. John Speed's plan from 1611. British King Edward I refounded the town and fortress of Caernarfon in July 1283 as part of his second expedition in Wales to quell the Second War of Independence.

Europe saw the construction of hundreds of new cities and the expansion of many more with well-planned additions between the ninth and fourteenth centuries. In the modern era, these newly established towns and town expansions have been instrumental in moulding the geographical structures of Europe. Although new cities began to spring up all throughout Europe beginning in the ninth century, the majority of these developments occurred during the twelve and fourteenth centuries, reaching a zenith towards the close of the thirteenth. Landlords of all classes sought to consolidate their economic, political, and military influence by colonising their estates with new settlements. The founding ruler either enticed or compelled people to relocate from his estates to the new cities because of the fiscal, economic, and legal benefits he offered. Most of the new towns were to remain rather small (as for instance the besides of South-western France), but some of them became important cities, such as Cardiff, Leeds, 's-Hertogenbosch, Montauban, Bilbao, Malmö, Lübeck, Munich, Berlin, Bern, Klagenfurt, Alessandria, Warsaw and Sarajevo [20].

Many of these settlements seems to have deliberately planned their formal architecture, according to the evidence of the surviving towns. The plan shape of the newly established towns is often rather regular; for example, the streets are typically straight and arranged at right angles to each other, and the home lots are rectangular and first quite uniform in size [21]. Elburg, in the Netherlands, which dates back to the late 14th century, is one plain and severe example. You can't possibly argue that the straight street and the symmetrical, orthogonal town layout were novelties of "the Renaissance" and, by extension, emblematic of "modern times" when you look at examples like Elburg's.

The era of tremendous urban growth came to a close with the terrible downturn that occurred in the middle of the fourteenth century. It would continue for another century or two only in regions of Europe where urbanisation had begun later, such Eastern Europe. Although the number of newly established communities would stay much smaller than in the 12th and 13th centuries, the same degree of urban population increase would not be achieved until the Industrial Revolution.

5 Renaissance and Baroque Europe (1400–1750)

A prime example of Renaissance urban planning aspirations is The Ideal City, likely created by Fra Carnival between 1480 and 1484. Victory in battle and the pleasure of a large audience are both symbolised by the Roman archway and the coliseum. Located in north-eastern Italy, the fortified town of Palma nova was constructed by the Venetian Republic in 1593. It is an example of a star fort from the late Renaissance. Florence served as a precursor to the new style of urban planning, which borrowed the star form of the fort to withstand cannon fire. Reflecting Florence's immense cultural prominence during this period, this style was extensively emulated. 'The Renaissance was hypnotised by one city form which for a century and a half - from Fila rete to Scamoglini - imprinted onto utopian schemes: this is the star-shaped metropolis" [22]. Radial streets extend outward from a defined centre of military, communal or spiritual power.

The ideal centrally planned urban space: *Sposalizio* by Raphael Sanzio, 1504

Centrally designed structures, like the one in Rafael's Sposalizio (Illustration) from 1504, were only found in perfect towns. As built, the unique instance of a rationally planned quattrocento new city centre, that of Vigevano (1493–95), resembles a closed space instead, enclosed by arcading.

The twelve-pointed design of the ideal city that filtrate created, which was based on Leon Battista Alberti's De re aedificatoria, was given the name "Sforzinda" as a tribute to his patron. The city ignored the undulating topography in the book, since it could be circumscribed by a "perfect" Pythagorean figure, the circle [23]. The industrial suburbs of the time (for more on this, see Braudel's The Structures of Everyday Life) remained chaotic, with overcrowding and organic expansion being defining features; yet, this transformation did take place in urban areas.

During the Baroque Era, Rome had one of the most ambitious city planning schemes. Pope Sixtus V's ambitious urban reform programme reshaped the old environment to resemble the Renaissance city ideal: "long straight streets, wide regular spaces, uniformity and repetitiveness of structures, lavish use of commemorative and ornamental elements, and maximum visibility from both linear and circular perspective." Through this programme, the old environment was transformed. The Pope set no limit to his plans, and achieved much in his short pontificate, always carried through at top speed: the completion of the dome of St. Peter's; the loggia of SIXT us in the Basilica di San Giovanni in Laterano; the chapel of the Praesepe in Santa Maria Maggiore; additions or repairs to the Quirinal, Lateran and Vatican palaces; the erection of four obelisks, including that in Saint Peter's Square; the opening of six streets; the restoration of the aqueduct of Septimius Severus ("Acqua Felice"); the integration of the Leonine City in Rome as XIV Rione (Borgo).

In an effort to showcase the splendour of their country, monarchs of this era often undertook grandiose endeavours to remodel their capital cities. In many cases, disasters served as a primary impetus for prepared rebuilding. However, this did not apply in London during the Great Fire of 1666, when architects like John Evelyn and Christopher Wren proposed several ambitious plans for reconstruction, but the intricacies of competing claims to ownership prevented any widespread remodelling. Wider streets, stone building, and river access, however, improved sanitation and fire safety. On the other hand, city planners in North America were prompted to re-evaluate urban planning after the Great Fire. Created in the wake of the Great Fire, the Grand Model for the Province of Carolina served as a blueprint for colonial administration. The Grand Model served as a starting point for the renowned Oglethorpe Plan for Savannah (1733).

A substantial portion of the city core was destroyed in the 1695 assault of Brussels by the French soldiers of King Louis XIV. Governor Max Emanuel suggested using the restoration to alter the city's layout and architectural style. Inspired by Turin, which was one of the first Baroque towns when Charles Emmanuel I, Duke of Savoy restored it beginning in 1600, his goal was to convert the mediaeval city into a Baroque metropolis [24]. Straight avenues, with its rational street plan, provided lengthy, unbroken vistas framed by similarly sized buildings. Locals and government officials were against the concept because they wanted things rebuilt quickly, had little money for extravagant ideas, and hated the idea of a new, alien type of architecture. Reconstruction efforts maintained the city's fundamental layout, although it differed from its pre-cataclysm state. Even though there wasn't enough money and the city needed to be rebuilt quickly, officials did all they could to make things better in terms of traffic, cleanliness, and aesthetics. To accommodate more traffic, some roadways were widened.

6 Enlightenment, Europe and America (1700–1800)

Despite the Great Fire of London, King Joseph I of Portugal and his ministers wasted little time in beginning the reconstruction of Lisbon after the 1755 earthquake. Boldly proposing to demolish major areas of the city and "laid out new avenues without limitation" was architect Manuel da Maia. The monarch and his minister went with the final choice even though it prioritised business and industry above royal and religious institutions [25]. Big squares, rectilinear, huge avenues, and expanded streets were the new mottos of Lisbon, commissioned by the eager monarch to create a city that was both fresh and properly arranged. One of the first examples of seismic protection in Europe may be seen in the Pomaline buildings [26].

Another significant urban plan from the Enlightenment period was that of Edinburgh's New Town, [27] built in stages between 1767 and 1850. The commercial and professional classes that resided in Edinburgh during the Age of Enlightenment were not pleased with the city's antiquated architecture. In January 1766, a contest was organised to discover a contemporary enough plan for the new neighbourhood.

The winner, James Craig, a 26-year-old architect, suggested a basic axial grid with a main road running down the crest that would connect two garden squares. Originally planned as a mostly residential suburb, The New Town would have included a variety of domestically styled professional offices. The site's economic potential was quickly recognised, despite the fact that there were few planned retail ground floors.

7 Modern Urban Planning (1800–onwards)

Urban planning became more complicated and specialised as a field of law and technology beginning in the 18th century. Among the first London planned buildings was Regent Street. Sir Christopher Wren and John Evelyn had started planning the city's reconstruction on the classical formal model shortly after the Great Fire of London (1666), so an organised system of streets could be established to replace the old arrangement. John Nash, who had been an advisor to the Prince Regent and had been assigned to the Office of Woods and Forests in 1806, and developer James Burton collaborated on the street's design. An Act of Parliament approved the plan in 1813, allowing the commissioners to borrow £600,000 for the construction and development process.

Development of London in 1800

It was anticipated that private capital would provide the majority of the street's revenue due to its commercial function. All property valuations and designs were

Nash's responsibility. Destroying several homes, stopping commerce, and contaminating the air with dust were all necessary steps in building the road. Leases on the new premises were first offered to current tenants.

The renovation in Paris was far more extensive. After being commissioned in 1852 to redesign the city's mediaeval street layout, Georges-Eugène Haussmann tore up large sections of the old quarters and laid out huge boulevards that stretched beyond the ancient city boundaries. The whole of Paris and its suburbs were part of Haussmann's master plan, which included rules for everything from building facades to public parks to sewers and waterworks to municipal utilities and monuments. The broad roadways had a purpose beyond mere sanitation and aesthetics: they made military mobility and police enforcement easier [28].

A comprehensive study of the city and its contemporary needs formed the basis of a parallel plan to expand Barcelona. After the city walls were destroyed in 1854, the Catalan engineer Ildefons Cerdà drew up plans to fill the area beyond them. He is said to have coined the word 'urbanisation' and articulated his concept in his 1867 work, General Theory of Urbanisation. A network of three broader avenues ran through Cerdà's Eixample, which means "extension" in Catalan. The neighborhood's 550 regular blocks had chamfered corners to provide room for trams. The blocks were oriented NW-SE to optimise sunshine and help with social interaction, with the goal of improving the residents' health. They were constructed around central gardens [29].

From 1857 onwards, ideas for Vienna's Ringstrasse were also being considered simultaneously. The city walls were replaced by this magnificent avenue. The city walls and moats were ordered to be demolished in 1857 by the command of Austrian Emperor Franz Joseph I. Numerous lavish public and private structures were constructed in the years that followed. Also in 1862, Berlin completed its "Bebauungsplan der Umgebungen Berlins" (Binding Land-Use Plan for the Environs of Berlin), which was meant to be valid for about 50 years. A vast perimeter was included in the proposal, which also included the territory around Berlin and Charlottenburg. Large swaths of dense urban city blocks, or "blockrand structures," were the end product of the concept. These buildings, which included both residential and commercial space, typically extended to the street and had a shared courtyard, which was eventually expanded by more court buildings to accommodate even more people.

As the century shifted into the twentieth, planning and architecture persisted in their paradigm change. Private companies in the 19th century generally drove the design and speed of construction in the industrialised cities, which saw significant growth during that time. More and more, the public was starting to notice the problems that the working poor faced in metropolitan areas. Urban planning as a mainly aesthetic and technical concern, as in the major urban planning programmes in European cities, was beginning to give way to a New Liberalism that advocated intervention on behalf of the poor and disadvantaged, in contrast to the laissez-faire style of government economic management that was popular throughout the Victorian era. Theorists started creating models of urban planning about 1900 to help people deal with the effects of industrialization, particularly factory workers, by making cities better places to live.

Development in Paris during 1800

The historic Innere Stadt neighbourhood of Vienna, Austria is encircled by the Vienna Ring Road, a circular magnificent boulevard in German (Ringstraße, lit. ring road). Where the road now stands, you can still see the crisscross of routes that led to the ancient city walls and ramparts (glacis) that protected the area throughout the Middle Ages.

Zoning laws, along with other tools like compulsory purchase and land readjustment, were developed in Prussia and later spread to the United Kingdom, the United States, and Scandinavia. These laws allowed urban planners to legally divide cities into sections for different functions or to determine the size and shape of urban blocks [31]. Public health was cited as a rationale for keeping cities organized [32].

8 Garden City Movement

The idea of Arturo Soria's of the linear city model and the "linear city movement" The seminal 1902 graphic by Ebenezer Howard depicting the expansion of cities via garden city "off-shoots"

Initiating the garden city movement in 1898, Sir Ebenezer Howard was the first prominent British urban planning thinker. Plans like Cadbury's Bournville, Lever's Port Sunlight, and Chicago's Pullman, named for its namesake developer, George Pullman, served as inspiration for this. All of these communities offered a healthier alternative to city life for manufacturing workers by moving their workplaces out of the heart of the cities. Howard expanded on this success to create a nationwide movement. Economist Alfred Marshall's arguments from 1884, in which he stated that industries required a supply of labour that could theoretically be found anywhere, and that businesses should strive to raise the living standards of their employees because they pay a disproportionate share of the price for the unhealthy urban conditions that plague major cities, also had an impact on him [40].

Despite their utopian bending Howard's proposals were widely implemented throughout the following decades. Planned, autonomous villages bordered by parks, his garden towns were to have housing, businesses, and farms in equal and distinct portions. Howard wrote Garden Cities of Tomorrow, widely considered the first work in the field of urban planning, in 1898 after being influenced by the utopian novel Looking Backward and Henry George's Progress and Poverty [41]. Built on a circular design with open spaces, public parks, and six radial boulevards, 120 ft (37 m) wide, stretching from the centre, his idealised garden city would accommodate 32,000 people on a land of 6000 acres (2428 hectares). Once the garden city reached capacity, it could support itself without outside help and a second garden city could be built nearby. Howard envisioned a network of interconnected garden cities, with a centre city of 50,000 people serving as its satellite. These cities would be connected by train and road. [42]

He founded First Garden City, Ltd. in 1899 to create the first garden city at Letchworth, Hertfordshire [43]. If the garden city was able to make money via rentals or what Fishman terms "philanthropic land speculation," the investors would get interest on their money [44]. Howard sought to include working-class cooperatives, which had more than two million members, but he was unsuccessful in securing funding from them [45]. The First Garden City, Limited held a competition in 1904 to design Letchworth, a region 34 miles outside of London. Renowned architect and town planner Raymond Unwin and his colleague Richard Barry Parker were the winners [46]. The town was designed by Unwin and Parker in the heart of the Letchworth estate, encircled by Howard's expansive agricultural greenbelt. Both men shared Howard's belief that the working class should have access to better and more reasonably priced homes. But the architects disregarded Howard's symmetrical plan and substituted a more "organic" style.

In addition, Howard's concepts were also used to construct Welwyn Garden City in Hertfordshire. When he stepped down as Garden City Association chairman, Sir Frederic Osborn continued the movement's expansion into regional planning [48].

Suburbs of cities began to be planned according to the garden city concept. Founded by Henrietta Barnett and designed by Parker and Unwin, the Hampstead Garden Suburb was the first of its kind. Utopian principles behind the plan called for low-density living, expansive tree-lined roadways, and free access to forests and gardens for all socioeconomic groups.

The "Neighbourhood Unit" style of development emerged from the wildly successful Garden City movement in North America. Concern about pedestrian injuries caused by automobiles initially surfaced in the early 1900s, when vehicles were first allowed on city streets. Neighbourhood Units, which were first seen in Radburn, New Jersey, were a solution that redirected home orientation away from streets and onto public pathways. The area is designed with a school as its focal point, ensuring a safe route for children to go to school.

Raymond Unwin, a student of Howard's, proved that houses could be constructed quickly and cheaply while still meeting acceptable standards for gardens, family privacy, and interior spaces; his work was incorporated by the Tudor Walters Committee, which advocated for the construction of housing estates following World War I. Unlike Howard, Unwin advocated for smaller, more outlying "satellites" rather than larger, more ambitious "garden cities."

9 Modernism

Partizánske in Slovakia – an example of a typical planned industrial city founded in 1938 together with a shoemaking factory in which practically all adult inhabitants of the city were employed.

Climate of Slovakia in 1900

Development of Slovakia in 1900

A 1939 model envisions Berlin transformed into the Welthauptstadt Germania (World Capital Germania).the Seven Sisters in Moscow in the "Wedding Cake Style Buildings" with the "ring roads" in the city.

Urban planners started to consider modernist concepts in the 1920s. In 1922, the renowned modernist architect Le Corbusier unveiled his plan for a "Contemporary City" (Ville Contemporaine) with three million residents. The sixty-story cruciform towers, which were office structures with steel frames and massive glass curtain walls, were the focal point of this proposal. The rectangular green grounds that surrounded these towers were enormous. Located in the middle was a massive transportation hub that, on various levels, included bus and rail stations, highway crossings and, at its highest point, an airport. The fantastical idea that commercial planes would land in the space between the massive towers was Le Corbusier's. He praised the car as a mode of mobility and created separate pathways for pedestrians to use instead of streets. Smaller, zigzag apartment complexes with low stories, located distant from the street and surrounded by green space, housed the people as one travelled out from the centre towers. French politically engaged businesspeople, according to Le Corbusier, would reorganise society by using efficient Taylorist and Fordist techniques borrowed from American models [53].

Demolition of the majority of central Paris north of the Seine and replacement with sixty-story cruciform towers from the Contemporary City, set inside an orthogonal street grid and park-like green space—this was his Plan Voisin, which he displayed in 1925. Before releasing La Ville radieuse (The Radiant City) in 1935, Le Corbusier spent the 1930s refining and developing his theories on urbanisation. The most striking distinction between the two cities is that the Radiant City did away with class-based stratification and instead distributed homes according to family size rather than economic status [54]. Public housing developers in the US and Europe sometimes embraced Le Corbusier's philosophies.

Nadir Afonso, a painter and architect, was one of many famous people who followed Le Corbusier and adopted his aesthetic ideology.The architect himself created the design for Chandigarh in India, and famous proposals based on his ideas include Lúcio Costa's city plan of Brasília and František Lydie Gahura's industrial city of Zlín in the Czech Republic. The Russian post-Revolutionary philosophies of Futurism and Constructivism had a significant impact on Le Corbusier's ideas as well.

Sir Patrick Geddes was another influential thinker; he saw the rise of massive urban conurbations, identified the connection between social concerns and town planning, and realised the need of considering the regional environment. While he was a prisoner in Mandatory Palestine in 1927, he was hired to design Tel Aviv. Roughly forty blocks, each measuring roughly 150 square metres, made it up. The block's windmill layout of inner access roadways made it difficult for cars to pass through the little public garden that was located within. To take use of the sea air, the large blocks are laid up in an east-west commercial, gently sloping street plan that runs north-south. Updating the old fixed grid patterns in this way was easy and quick. An alternative route that bypasses the system is a network of shaded boulevards that go to the waterfront via a number of public plazas. It all worked out according to plan for the new community.

Many communist nations' urban planning practices are reminiscent of Western modernism, with central planning offices enforcing state-sponsored, highly efficient city plans. (While this procedure was ostensibly decentralised in Russia with the fall of the USSR, the majority of the country's urban planning knowledge is still based in Moscow) [55]. National socialism also led to Germany's ambitious plans to reshape its cities [56].

10 New Towns

The ideas put out by Ebenezer Howard for city planning did not see widespread adoption until after WWII. The war's devastation piqued the public's curiosity about Britain after the conflict, and the government actively promoted this by facilitating discussions on a "Better Britain" in an effort to raise morale. As part of post-war reconstruction efforts, new plans were drawn up for London that finally tackled the problem of centralization. Firstly, in order to rebuild the city at an acceptable density, it was acknowledged in the County of London Plan 1943 that population and employment relocation was inevitable. More than a million people would have to be relocated to a combination of new towns, satellite suburbs, and existing rural towns according to the Greater London Plan of 1944 [57].

A 0 series Shinkansen high-speed rail set in Tokyo, May 1967

The New Towns Act 1946 (9 & 10 Geo. 6. c. 68) resulted in many New Towns being constructed in Britain over the following decades.[58, 59]

The 1960s marked the start of a wave of new American cities such Riverside Plaza, Jonathan, Minnesota, Columbia, Maryland, and Reston, Virginia. Slum removal, improved and expanded housing, road development, and comprehensive urban rehabilitation projects were funded by the federal government, which matched these construction efforts. France, Germany, Italy, and Sweden were among the other European nations that had considerable success with new cities, particularly during the post-war rebuilding period.

Contemporary Urban planning has grown in prominence with rising urbanization [60].

11 Reaction Against Modernism

A great deal of planners in the late 1960s and early 1970s held the view that the community's energy was drained by modernism's impersonal size and clean lines, and they held this view as the root cause of social issues and high crime rates [61].

In the 1970s, when most nations, including France and Britain, stopped building inexpensive, identical tower blocks, modernist planning began to wane. Many have been torn down and rebuilt with other kinds of homes since then. This is the post-modernist period, where planning focuses on individuals and economic and social variety rather than trying to eradicate all chaos [61].

There are still cities with little planning. There is no comprehensive zoning code in the developed nation where the big city of Houston is located, despite its 5.5 million people in the metropolitan area. Despite the lack of regulation for certain land uses, Houston does limit building density and requires parking. Private developers in Houston also impose land-use regulations similar to zoning laws via the use of deed restrictions and subdivision covenants. Three times since 1948, voters in Houston have rejected comprehensive zoning regulations.

12 Behaviorism

Defensible space and crime prevention via environmental design are two examples of the ways in which behaviourist psychology impacted city planning, particularly in the 1960s and beyond [62].

13 New Urbanism

Jakriborg in Sweden, started in the late 1990s as a new urbanist eco-friendly new town near Malmö

Several contemporary schools of thought in urban planning aim to establish livable, long-lasting cities that are easy on the planet and its residents. The most well-established model of pedestrian-friendly city planning is the New Urbanism Charter. It is a strategy for establishing and maintaining smart cities that facilitate sustainable transit while simultaneously lowering their negative effects on the environment. Compared to those living in broad suburbs, residents of small urban areas travel less kilometres and have far lower environmental effects across a variety of metrics. . In an effort to encourage sustainable land use patterns that aim for compact cities and a decrease of greenfield land absorbed by urban sprawl, the idea of circular flow land use management has also been adopted in Europe.

Doha, Qatar, site of UNCTAD XIII, a road crossing. Two roadways, one for each direction, divide traffic in Qatar.

Smart expansion, walkability, architectural tradition, and classical architecture are all valued and fostered by the New Classical Architecture movement, which advocates for sustainable urban building. This goes against the grain of modernist, fleeting, universally identical architecture, as well as isolated housing developments and sprawling suburbs. Both trends started in the 1980s.

Some people think that the environmental aspects of New Urbanism are too concerned with transportation and people's tendency to move around a lot. Urban metabolism as a whole, not just cars and excessive driving, is to blame for today's cities' unsustainable character, according to these critics. While auto-mobility contributes to half of the city's greenhouse gas emissions and ecological footprint, it accounts for less than half of the city's total ecological footprint. They have also said that regulation of the design and related technologies of the actual development inside a designated region is necessary for land-use planning to achieve sustainability. They argue that overall development metabolism, rather than distance or density, is what defines environmental effect. In addition, they argue that the focus should change from sustainability to resilience, and that the area under consideration should expand beyond the city limits to include the whole metropolitan region. The fact that New Urbanist efforts to compress urban form are laborious and time-consuming is another point of contention. The authors argue that in light of the current global situation—characterized by climate change and the irreversible dominance of horizontal, low-density growth—it would be prudent to prioritise the resilience of entire city-regions. This can be achieved by transforming the current sprawl into a more sustainable and self-sufficient form, and by making substantial investments in "green infrastructure."

References

1. Dalley (1989), p 120
2. Davreu R (1978) 'Cities of mystery: the lost empire of the Indus valley. In: The World's Last Mysteries. (Second edition). Sydney: Reader's Digest. pp 121–129. ISBN 978-0-909486-61-7
3. Kipfer BA (2000) Encyclopedic Dictionary of Archaeology. (Illustrated edition). New York: Springer. p. 229. ISBN 978-0-3064-6158-3
4. Ben-Joseph E (2012) Codes and standards. In: Weber and Crane, p 357: Other excavations, such as those in the Indus Valley, at Mohenjo-Daro and at Harappa in Punjam, show that the valley had cities planned and built in rectangular blocks lined with two-story houses along broad, straight streets. Such evidence of the pervasive use of orthogonal arrangements tells of an enduring continuity of farm, village, and city. The farmer's custom of long narrow fields and right-angled boundaries carried easily to streets and squares. However, the innovative geometry of towns and cities was influenced more by the mechanics of supplying water and draining sewage than in the imposition of right-angled street corners
5. Smith, ME (2005) City size in late post-Classic Mesoamerica (PDF). J Urban History 31(4). Beverley Hills, CA: SAGE Publications: 403–434. https://doi.org/10.1177/009614420427 4396. OCLC 1798556. S2CID 145452272. Retrieved 1 Feb 2008
6. Jo-Bo System of Heian-kyo–City Planning in Classical Japan
7. Aristotle, Politics II
8. Hippodamian Plan–Livius
9. Paden R (2001) The two professions of Hippodamus of Miletus. Philos Geo 4(1):25–48. https://doi.org/10.1080/10903770124644. ISSN 1090-3771. S2CID 144920561
10. Urban planning in ancient Egypt
11. Kolb F (1984) Die Stadt im Altertum. München: Verlag C.H. Beck. pp 51–141: Morris AEJ (1972). History of Urban Form. Prehistory to the Renaissance. London. pp. 22–23
12. Hippodamus is also reported to have played an important role in the (re-)planning of the city of Piraeus, and he is also associated to Thurii and Rhodes. It seems that the importance of Hippodamus' ideas was not so much in the orthogonal plan, but rather in the distribution of functions and social classes across the city. (Kolb 1984, pp. 115-120; Cahill, Nicholas (2002). Household and City Organization at Olynthus. New Haven. par. 1.2)
13. Morris (1972) pp 39–41, 51–60; Kolb 1984, pp. 169–238; Benevolo, Leonardo (1993). Die Geschichte der Stadt. Frankfurt am Main/New York. pp 256–267
14. Harris W (1989) Invisible cities: the beginning of Etruscan Urbanization. Atti del Secondo Congresso Internazionale Etrusco. Roma, 1989. pp 375–392. p 85. The Etruscans were, in their turn, probably also influenced in this respect by Greek and Hellenic culture
15. Jerke P, Lassar (2008) pp 8–9. In: His [Mark Salette of Gehry Partners] view, the core qualities of good design—loosely based on principles laid down by Vitruvius in the first century B.C. and countless reconceptualizations since—generally are functionality and durability, contextual compatibility, and enduring respect and value
16. Vitrivius (1914) The ten books on architecture, Bk I. Morris H. Morgan (translator). Harvard University Press
17. Demandt A (1998) Die Kelten. München: Verlag Ch. Beck. In fact, many sites where the Romans created towns, such as Paris, Vienna and Bratislava, had previously been Celtic settlements of more or less urban character

18. Bartlett R (1993) The making of Europe. Conquest, Colonization and Cultural Change 950–1350. London/New York. pp 5–60, 106–197; Gutkind, A.E. (1964) International History of City Development. Vol.I: Urban Development in Central-Europe. New York/London. pp 13–20, 63
19. Siegfried Giedion, Space, Time and Architecture (1941) 1962, in reference to an air view (Fig. 8) of the medieval Italian town of Bagnocavallo. Giedion's source was Luigi Piccinati, "Urbanistica Medioevale" in Urbanistica deal Antichità ad Oggi (Florence 1943)
20. Boerefijn W (2010) The foundation, planning and building of new towns in the 13th and 14th centuries in Europe. An architectural-historical research into urban form and its creation. Phd. thesis Universiteit van Amsterdam. ISBN 978-90-9025157-8 ([1])
21. Beresford, Maurice (1967) New Towns of the Middle Ages. Town Plantation in England, Wales and Gascony. London ; Gutkind, E.A. (1972) International History of City Development. Vol.VII: Urban Development in East-Central Europe: Poland, Czechoslovakia and Hungary. New York/Londen; Lavedan, Pierre & Jeanne Hugueney (1974) L'Urbanisme au Moyen Age. Geneva/Paris; Friedman, David (1988) Florentine New Towns. Urban Design in the late Middle Ages. New York/Cambridge (Mass.)/London; Lauret, Alain, Raymond Malebranche & Gilles Séraphin (1988) Bastides, villes nouvelles du moyen-age. Toulouse; Guidoni, Enrico (1992) Storia dell'urbanistica. Il Duecento. Roma/Bari
22. Siegfried Giedion, Space, Time and Architecture (1941) 1962 p 43
23. The undulating terrace of housing makes its appearance surprisingly late: Giedion's example is Lansdown Crescent, Bath, 1794; Giedion 1962, Fig. 83
24. Klaiber S (1993) Review of Turin, 1564–1680: Urban design, military culture, and the creation of the absolutist capital; military architecture, cartography, and the representation of the early modern European City: a checklist of treatises on fortification in the Newberry library, Martha d. Pollak. J Soc Archit Hist 52(1):101–103. https://doi.org/10.2307/990765.ISSN0037-9808. JSTOR990765
25. Shrady, The last day. pp 152–155
26. Pombaline Lisbon. UNESCO World Heritage Centre. Retrieved 26 Oct 2021
27. Old and new towns of Edinburgh. UNESCO World Heritage Centre. Retrieved 26 Oct 2021
28. Girouard M (1985) Cities and people, London, p 285
29. Busquets J (2005) Barcelona, the Urban evolution of a compact city, ISBN 978-88-8447-204-5, Harvard University, p 122
30. Larsson G (1997) Land readjustment: a tool for Urban development. Habitat Int 21(2):141–152. https://doi.org/10.1016/S0197-3975(96)00059-8. ISSN 0197-3975
31. Ben-Joseph E (2012) Codes and Standards. In: Weber & Crane, p 359: In their search for a scientific model for the American city, these newly formed experts looked overseas. According to Frederick Law Olmsted, Jr. (1870–1957), the president of the National Planning Conference and the first president of the American City Planning Institute (1917), the model was found in Prussia (Germany). After the Franco-Prussian war (1870–1871), many Prussian cities experienced uncontrolled buildup of inner-city factories and housing, followed by dangerous health hazards and pollution. Citizens protested the deteriorating environmental conditions. Thereupon, Prussian city officials responded with a zone system that assigned particular land uses to designated locations
32. Corburn J (2012) Reconnecting Urban planning and public health. In: Weber & Crane, p 398: Consistent with the emerging aesthetic and technocratic ideals of planning, tools to classify and segment city functions took hold and land-use zoning emerged as one such powerful tool. Local zoning ordinances were often couched as both protecting public health and benefiting private landowners. For example, in the landmark 1926 Supreme Court case *Village of Euclid, Ohio, et al. v. Ambler Reality Company*, zoning was characterized as promoting health: 'the exclusion of buildings devoted to business, trade, etc., from residential districts, bears a rational relation to the health and safety of the community...by excluding from residential areas the confusion and danger of fire, contagion and disorder, which in greater or less degree attach to the location of stores, shops, and factories' (272 U.S. 365, 392). Yet, there was little to no epidemiologic evidence at the time suggesting that health improvements would come about by physically separating land uses, and some have suggested that zoning tended to concentrate health hazards and perpetuate disadvantages in poor urban communities (Maantay 2001)

33. Sliwka R (2011) Toronto's waterfront: hybrid landscapes and infrastructures of desire. Magazine of Sustainable Design (Quadrimestrale on line sul progetto di città sostenibile) Edizione SCUT, Università Chieti-Pescara. **7**
34. Stubben J (2008) City building (Der Stadtebau)–English Translate. MIT Roch Library. pp. Forward–Emily Talen and Julia Koschinsky
35. Franck HH (1921) Die Verwertung von synthetischen Fettsäureestern als Kunstspeisefette. https://doi.org/10.1007/978-3-663-07056-6. ISBN 978-3-663-06143-4
36. Hall T (2003) Planning Europe's capital cities. Routledge. p 389. ISBN 978-0-203-44956-1
37. Urban planning
38. The birth of town planning
39. RTPI.org.uk, About the RTPI. Retrieved 17 Jan 2013
40. Hall P, Tewdwr-Jones M (2010) Urban and regional planning. Routledge. ISBN 978-0-203-86142-4
41. To-morrow: a peaceful path to real reform
42. Goodall B (1987) Dictionary of human geography. Penguin, London
43. Hardy (1999), p 4
44. Fainstein and Campbell 2003, p 43
45. Fainstein & Campbell (2003), p 46
46. Hall (2002) p 68
47. Fainstein and Campbell (2003) p 48
48. History 1899–1999 (PDF), TCPA, archived from the original (PDF) on 27 July 2011
49. Gayler, HJ (1996) Geographical excursions in London. University Press of America. p 176. ISBN 978-0-7618-0328-7. Retrieved 24 Aug 2011
50. Christensen CA (1986) The American garden city and the new towns movement
51. Schaffer D (1982) Garden cities for America: the Radburn experience (Temple University Press 1982)
52. Hall P (2002) Cities of tomorrow. Oxford: Blackwell. p 74. ISBN 978-0-631-23252-0
53. Evenson N (1969) Le Corbusier: The Machine and the Grand Design. George Braziller, Pub: New York, p 7
54. Fishman R (1982) Urban utopias in the twentieth century: Ebenezer Howard, frank Lloyd wright, and le Corbusier. MIT Press, Cambridge, MA, p 231
55. Friedman J (2012) Varieties of planning experience: toward a globalized planning culture?. In: Weber & Crane
56. Cuthbert (2006), p 56: Rationalist architecture and urbanism reached its zenith in its coincidence with functionalist social science and the eugenic strategies of fascism, particularly with Hitler and his architect Albert Speer, and with Mussolini (Marcello Piacentini, Giuseppi Terragni)
57. Osborn J, Whittick A (1977) New towns: their origins, achievements, and progress. Third Ed. Leonard Hill, London. First Pub 1963. pp 55
58. Osborn J, Whittick A (1977) New towns: their origins, achievements, and progress. Third Ed. Leonard Hill, London. First Pub 1963, pp 56
59. Hall P, Ward C (1998) Sociable cities: the legacy of Ebenezer howard. John Wiley and Sons, Chichester, pp 41–69
60. Weber R, Crane R (2012) Planning as scholarship: origins and prospects. In: Weber & Crane
61. Planning SMBT, Design U et al (1997) ISBN 978-0-582-23496-3. Longman, Singapore
62. Xu L, Du LS, Wang C, Xu W (2012) Nanofiltration coupled with electrolytic oxidation in treating simulated dye wastewater J Membr Sci 409(2012):329-334 View PDFView articleView in ScopusGoogle Scholar

Sustainable Development and Sustainability

Biman Gati Gupta

Sustainable development has emerged in recent decades as guiding themes for urban planning. This term was defined and advocated in 1987 report Our Common Future, published by the World Commission on Environment and Development.

Robert Hoddle's survey of Melbourne in 1837. The layout of the city is referred to as the "Hoddle Grid".

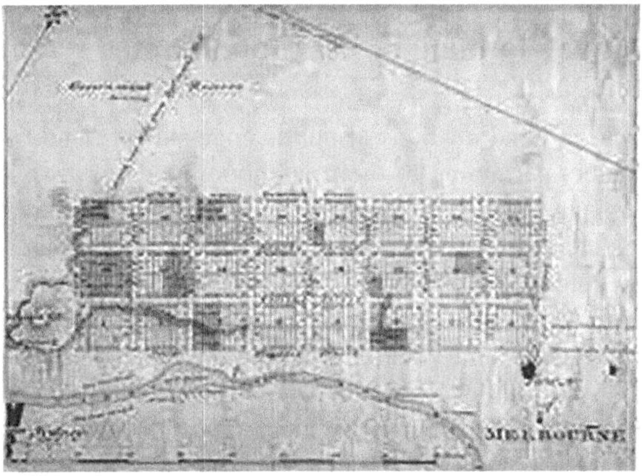

Robert Hoddle's survey of Melbourne in 1837. (Hoddle's Grid)

Some urban planners contend that contemporary ways of living are to blame for a number of problems, including the depletion of natural resources, the worsening of social inequality, the development of urban heat islands, and the acceleration of global warming. As a result, sustainable cities are a popular concept among urban planners.

B. G. Gupta (✉)
Elitte College of Engineering, MAKA University of Technology Sodepur, Kolkata 700113, India
e-mail: bimangupta@yahoo.in

B. G. Gupta (ed.), *Urban Development with Treated Recycled and Reuse of Wastewater: Sustainable Management*, Urban Sustainability,
https://doi.org/10.1007/978-981-96-0857-7_2

But the idea of sustainable development is very new and has generated some debate. SUD is "development that enhances the long-term social and ecological health of cities and towns," according to Wheeler's 2004 book. He outlines the characteristics of a "sustainable" city, which include: efficient and compact land use; less reliance on automobiles while improving accessibility; efficient use of resources; less pollution and waste; restoration of natural systems; decent housing and living conditions; a strong social ecology; a sustainable economy; community engagement and involvement; and the preservation of local knowledge and traditions. A sustainable city model is being advocated by urban planners. This model calls for cities to be built in a way that minimises their environmental consequences, including energy and water consumption as well as waste and pollution outputs.

Sustainable planning initiatives need widespread support before they can influence institutions and regions, due to the political and governance frameworks in most jurisdictions. A difficult compromise is typically required for actual execution.

Nature in cities Often an integral party of sustainable cities is the Incorporation of nature within a city.

Car free sustainability in city planning can include large pedestrian zones or be a totally Car free.

1 Collaborative Planning in the United States

The failure of conventional methods of public engagement to allow citizens meaningful say in matters pertaining to their neighbourhoods gave rise to the concept of collaborative planning in the United States. In order to encourage participation, ideas, and engagement in the community planning process, collaborative planning seeks to put stakeholders in a position of power by bringing them into the decision-making process via direct interaction and communication between stakeholders and public authorities. Planners may benefit from public participation in two ways: first, by learning about the public's wants and requirements; and second, by incorporating local expertise into projects. More meaningful engagement and better, more innovative solutions to long-lasting issues may be achieved via well-managed collaboration compared to more conventional participation techniques. By include community members, planners may make choices that are in line with community values and needs, which in turn increases community trust in the project's wisdom and practicality and gives everyone a personal investment in its success.

The success of collaborative planning in Portland and Seattle hinges on a few interconnected factors. First, it must be truly inclusive, meaning that all affected groups and stakeholders are invited. Second, the community must have the last say in decision-making. Third, the government must clearly commit financially and intellectually. Fourth, planning staff should provide participants with clear objectives and act as facilitators, offering advice, research, and expert opinions. "Finally, facilitators should have training in community organisation and conflict resolution.

2 Urban Planning

The primary focus of urban planning, an expansive discipline, is the formulation of all-encompassing plans and policies for the improvement and expansion of urban areas and surrounding regions. Urban planners aim to create communities that are both functional and environmentally sustainable by balancing issues including population growth, land use, transit, and sustainability.

Examples of urban planning initiatives include:

- the development of zoning regulations,
- land use plans,
- transportation networks,
- affordable housing policies.

For example, urban planning encompasses the process by which a city decides which areas will be used for residential, commercial, or industrial reasons. The field's overarching goal is to create a plan for the city's growth that will bring its many parts into harmony.

3 Details of Urban Design

The field of urban design, in contrast, focuses on the practicality, aesthetics, and human experience of built metropolitan environments. Urban planning is the practice of creating places that are aesthetically pleasing, functionally sound, and culturally significant by the strategic placement and design of buildings, roadways, and public spaces.

Examples of urban design projects include:

- the layout of public squares,
- the design of pedestrian-friendly streetscapes,
- the revitalisation and regeneration of historic districts.

Spaces that are both practical and visually beautiful, while also fitting in with the area's history and culture, are typically the goal of urban designers. Their job has the potential to significantly shape how a city or community is seen and understood.

4 Urban Planning and Urban Design

When it comes to modernising and improving cities and the amenities they provide, urban design and urban planning work hand in hand. The main distinctions between them, nevertheless, are as follows:

Scope: When it comes to city development, urban planning is all-encompassing and strategic, looking at policies and objectives for the future, while urban design is more concerned with the layout and aesthetics of individual spaces or projects.

Scale: While urban design focuses on more intimate spaces and projects, urban planning considers bigger areas, such as whole cities or regions.

Objectives: The goal of urban design is to improve the aesthetic and experiential aspects of urban places, while the goal of urban planning is to build sustainable and functioning cities via legislation and regulation.

Timeframe: Results from urban planning projects don't often become apparent for years or even decades. Timeliness of execution may be less of a concern for urban design initiatives.

Urban Planning

Urban planning is the process of creating and controlling places for various purposes within an urban setting with an eye towards the built form, economic functions, and social implications of such places. Urban planning is at once a technical profession, an activity requiring political will and public engagement, and an academic field due to the fact that it relies upon engineering, architecture, and social and political considerations. Urban planning encompasses a wide range of activities, including but not limited to: public engagement, goal formulation, data analysis, forecasting, design, and the creation of both new areas of land (''greenfields sites") and existing urban areas that need rejuvenation. Geographic information systems (GIS) have become more popular tools for mapping the current urban system and predicting how changes will affect it. The phrase ''sustainable development" emerged in the late twentieth century to describe the pinnacle of planning aspirations. ''Development that satisfies the demands of the present without compromising the capacity of future generations to meet their own needs.'' This is the definition of sustainability put forward by the UN-sponsored World Commission on Environment and Development in their 1987 report, Our Common Future. Despite the fact that everyone agrees on the big picture, conflicts arise regularly since most important planning choices entail sacrificing one or more secondary goals.

Urban planning

In response to the chaos of the industrial city, a social movement for urban reform emerged in the late nineteenth century; this movement laid the groundwork for contemporary urban planning. The urge for planning was driven by both idealistic visions of the perfect city and more realistic concerns about the supply of facilities, the mobility of people and commodities, and appropriate sanitation. Social justice, economic development, environmental consciousness, and aesthetic appeal are among competing goals that modern planners attempt to meet. The end product of a planning process may range from a comprehensive strategy for a whole city or metropolitan region to detailed blueprints for individual neighbourhoods, projects, or even a collection of policy options. Even while planners and their sponsors try to keep politics out of planning, successful plan execution typically needs entrepreneurial spirit and political acumen. The private sector is playing an ever larger role in government-based planning via the formation of "public–private partnerships".

The early twentieth century saw the birth of urban planning as an academic field. In 1909, the University of Liverpool in Great Britain launched the first academic planning programme. In 1924, Harvard University in the United States developed the first programme of its kind. Its curriculum differs substantially from one institution to another, and it is mostly taught at the doctoral level. Programmes that provide doctorate degrees tend to focus on the social sciences, while others stick to the tried-and-true methods of physical design and land use. The problems it seeks to solve, rather than any prescriptive technique or dominating paradigm, provide a clearer definition of the theoretical core of the field. Some examples of representative issues include: the process for identifying and determining a public interest, the ideal city's physical and social characteristics, the likelihood of accomplishing change in line with goals that have been intentionally set, the communication's ability to bring about consensus on goals, the relative contributions of citizens, public officials, and private investors to city planning, and, more specifically, methodological concerns about quantitative analysis and the Environmental policy, transit planning, housing, and neighbourhood economic development are just a few of the many practical subjects included in urban planning degree programmes.

Urban design

5 The Development of Urban Planning

Early history

The remains of ancient towns in South and Central America, Asia Minor, Egypt, India, and the Mediterranean region have shown indicators of meticulous design. Efforts towards planned urban development can be seen in the following early examples: rectilinear and occasionally radial street systems; city division into specialised functional quarters; development of commanding central sites for civic buildings, temples, and palaces; and advanced systems of fortification, water supply, and drainage. The majority of the proof may be found in tiny towns that were established as colonies in very little timespan. The capital cities of ancient nations sometimes expanded to considerable sizes before they established administrations that could exert authority.

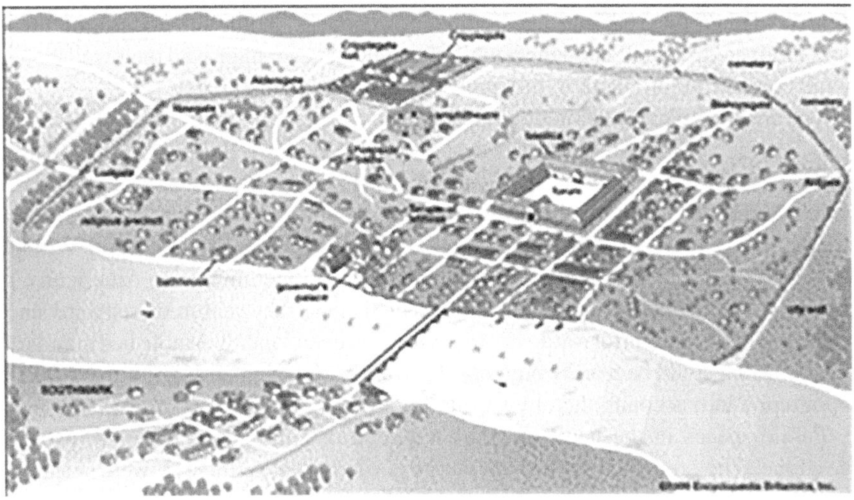

Roman settlement of londinium

The Roman settlement of Londinium, *c.* AD 200, which developed into the modern metropolis of London (more).

During the Middle Ages, construction of European towns remained minimal for a number of centuries. Churches, feudal authorities, marketing hubs, and trading posts emerged as towns evolved throughout time. Overcrowding, air and light pollution, and very bad sanitation were the results of urban population growth and the constrictions imposed by walls and fortifications. Like in many modern cities in the developing world, certain areas of cities were and are off-limits to people of certain ethnicities, socioeconomic groups, or occupations.

Cityscapes throughout the Middle Ages and the Renaissance tended to be more circular or irregular in shape, following the layout of villages along streets and crossroads; nevertheless, some of the more recent towns tended to be more rectangular in shape. Even in large European towns, pavement was not generally implemented until the twelfth century (1184 in Paris, 1235 in Florence, and 1300 in Lübeck). Prior to this, most streets served primarily as pathways, serving more as a means of communication than for transportation. Walls were often extended to accommodate the expanding populations of cities, but at that period, very few cities had walls longer than one mile. Sites were sometimes relocated, like in Lübeck, and several new cities sprang up in response to population growth, often within a day's walk of one other. Populations in towns varied from a few hundred to as high as forty thousand (like London in the late fourteenth century; although, London's population had reached eighty thousand prior to the Black Death). With 100,000, Paris and Venice stood out.

The Renaissance saw a renaissance in European efforts to consciously design towns. The glorification of a ruler or state was generally the primary goal, even while these endeavours attempted to improve circulation and provide military defence to some extent. Numerous towns were planned and constructed with colossal magnificence between the 16th and late eighteenth centuries. While the end product may have motivated and delighted the populace, it almost never improved their well-being, home comfort, or production, distribution, or marketing efficiency.

The absolute planning principles of Europe only partially found their way into the New World. This transference was best shown by Pierre L'Enfant's (1791) extravagant design for Washington, D.C., and by subsequent City Beautiful initiatives, which sought to maximise the beauty of public building placement at the expense of practicality in residential, commercial, and industrial development. But William Penn's (1682) strict grid plan of Philadelphia had a greater impact on American city planning. The pioneers carried this design with them as they ventured westward since it was the most straightforward way to divide surveyed land. It established standard-sized lots that could be readily purchased and sold, even sight unseen, without taking topography into account, thereby facilitating the growth of land markets.

Public spaces in the heart of cities were the original notion in many parts of the globe. The strategies for residential construction, nevertheless, were different. American towns took after their New England counterparts in the layout of its centre commons, which served as a pasture in the beginning but now hosts a variety of community institutions including a meetinghouse, tavern, smithy, and stores. The freestanding single-family dwelling, which became the standard in most urban areas, also originated in the New England town. Even in European city layouts, the centre plaza, piazza, or square served as the focal point. In contrast to American residential growth, the attached home was the dominant form of domestic architecture in Europe. In other parts of the globe, markets or bazaars, not open spaces, were the focal point of city planning. Many African and Asian communities were composed of modest cottages gated off from the street, whereas the Mediterranean area was characterised by courtyard-style domiciles.

The Era of Industrialization

Biman Gati Gupta

Jacob Riis: photograph of a New York City tenement

Shelter for immigrants in a New York City tenement, photograph by Jacob Riis, 1888 (more).

During the industrial revolution of the mid- to late-nineteenth century, both the United States and Europe saw unprecedented rates of population increase, unbridled corporate greed, massive speculative gains, and a general inability to control the negative physical effects of progress. During this time, massive urban centres sprung up, displaying the extremes of riches and poverty side by side. City planning was an aspect of the Progressive movement that emerged in response to the exploitation and

B. G. Gupta (✉)
Elitte College of Engineering, MAKA University of Technology Sodepur, Kolkata 700113, India
e-mail: bimangupta@yahoo.in

B. G. Gupta (ed.), *Urban Development with Treated Recycled and Reuse of Wastewater: Sustainable Management*, Urban Sustainability,
https://doi.org/10.1007/978-981-96-0857-7_3

corruption of the time. People reacted by demanding better sanitation as a means to combat the slums, congestion, chaos, unsightliness, and illness danger. Essential to the continued expansion of urban populations, technical advances to water supply and sewerage led to a marked improvement in public health. Earlier reforms to the housing market were passed later in the century. Construction standards for housing were first defined by early regulatory legislation, such as the Public Health Act of 1848 in Great Britain and the New York State Tenement House Act of 1879. But implementation was sluggish since neither governments nor slum residents could afford to upgrade old homes, and landlords had little motivation to do so due to the poor incomes of slum inhabitants. Still, housing became better as more buildings were up and laws were passed to make things better, sometimes in reaction to investigations and campaigns led by people like Charles Booth in England and Jacob Riis in the US.

Mall at Central Park

The Mall at Central Park, New York City, in 1902.

Recognising the need for leisure also led to initiatives to enhance the urban environment throughout the Progressive period, which lasted into the early twentieth century. Parks were created to provide a visual respite and areas for active recreation or unwinding. Congested spaces were later turned into playgrounds, and facilities for games and sports were built for both children and adults, whose workdays were increasingly decreased. The working class was relegated to unhealthy dwellings and congested jobs, but advocates of parks hoped that access to outdoor leisure would help them become more civilised. Architects Calvert Vaux and Frederick Law Olmsted had a vision for New York City's Central Park in the 1850s, and their work became an inspiration for many. Notable achievements included dividing pedestrian and vehicular traffic, transforming an urban core into a picturesque setting, and showing how parks may significantly boost local property prices. (Refer to landscape landscaping).

Throughout Europe, the imperial heritage of palaces and courts, as well as the key plazas and magnificent structures of church and state, had long shown concern for the city's look. During the Second Empire (1852–171) in Paris, Georges-Eugène, Baron Haussmann became the most prominent large-scale designer by promoting symmetrical layouts with radiating roads and squares, as well as straight arterial boulevards and excellent views. All throughout continental Europe, people began modelling their city plans after the one that emerged. Haussmann's modernization of the city allowed for the efficient transit of products and the speedy mobilisation of armed soldiers, going much beyond mere beauty. They effectively broke down the obstacles to trade that mediaeval Paris had offered. The eviction of low-income residents from centrally located areas, the construction of transportation corridors and commercial space that divided residential neighbourhoods, and the replacement of older tenement buildings with newer, more affluent apartment houses were all part of his plans. From the early twentieth century until the late 20th, urban rehabilitation programmes in the US and Europe followed Haussmann's model; after that, they spread to much of the developing globe.

The World's Columbian Exposition of 1893 in Chicago, designed by American architect Daniel Burnham, was the crown jewel of the City Beautiful movement, which popularised the opulence of European architecture in America. The exposition's architectural style set a standard that several towns tried to match. Civic centres and boulevards throughout the nation began to take design cues from the Greco-Roman ideal of the City Beautiful, with its expansive plazas and magnificently placed public buildings, in an effort to stand in stark contrast to the chaos and ugly that pervaded their environs. Unfortunately, the City Beautiful model's little capacity to boost companies' profitability and the far smaller authority of the state in the US both stifled its dissemination.

The utopian idea of the garden city, originally outlined by British social reformer Ebenezer Howard in his book Garden Cities of To-Morrow (1902), impacted the look of residential areas in the US and UK, whereas Haussmann's approach had a greater impact on Europe and the design of American civic centres. Basically a suburban layout, Howard's garden city had low-rise houses on meandering streets and cul-de-sacs, commercial and residential areas were kept apart, and there was an abundance of green space. Howard advocated for a "cooperative commonwealth" whereby residents would all benefit from increased property prices, social ownership of open space, and close proximity of businesses to homes. The successors to Howard rejected his socialist principles but continued to follow the pattern of residential design seen in the two new towns that were established during his lifetime, Welwyn Garden City and Letchworth. Substantially, the modern suburban subdivision was modelled after the garden city with its meandering roads and abundant greenery.

Modern transport systems are perhaps the most important contributor to the built environment of modern cities. Superhuman urban growth was made possible by the transition from foot and horse travel to the use of mechanical vehicles. Goods could travel swiftly from the site of production to the market, and workers may dwell far from their places of employment. The older sections of cities, however, saw a significant increase in traffic congestion due to cars and buses. The need to create new forms of organised circulation networks was emphasised by the prospect of traffic suffocation. With the advent of the metro systems in Paris, London, and New York City at the turn of the twentieth century, planning efforts shifted their attention to transportation networks. Local governments spent a lot of money expanding and upgrading roads to handle the influx of vehicles. Check out traffic control as well.

The first three decades of the twentieth century saw the establishment of planning departments by several local governments. In 1909, a number of important events occurred that would later come to define urban planning as a modern governmental function. These included the first town-planning act in Britain and the first national conference on city planning in the US. Additionally, Burnham's plan for Chicago was published and the Chicago Plan Commission was appointed. It's worth noting that the first recognised planning agency in the US had been established in Hartford, Connecticut, in 1907. Around the same time, other European nations like Germany and Sweden advanced their own systems of planning, administration, and law.

New Delhi, India: Presidential House (Rashtrapati Bhavan)

The Presidential House (Rashtrapati Bhavan), formerly the Viceroy's House, New Delhi, India, designed by Sir Edwin Lutyens, constructed 1913–30 (more).

European ideas of city planning were brought to emerging global cities by the colonial powers. A new city, designed in accordance with Western ideals of aesthetics and functional segregation, would frequently rise up next to older, less planned communities, enduring all the problems of a mediaeval European metropolis. Delhi, India, is a prime example of this kind of growth. It sprang up side by side with the maze of Old Delhi streets, built according to a design by British architects Edwin Lutyens and Herbert Baker. In contrast, the less pristine ancient city provided its residents with a feeling of belonging, continuity with their past, and practicality that was better suited to their daily lives. Capitals of African countries under British administration, such Nairobi, Kenya, and Salisbury, Southern Rhodesia (now Harare, Zimbabwe), followed a similar pattern to satisfy the white colonial rulers. French planners also installed wide boulevards and European-style houses in its colonial colonies, but the ornamental themes enforced by France in its colonial capitals reflected a somewhat different aesthetic sense.

1 Urban Form

Zoning and subdivision controls

In the early twentieth century, as Western industrial towns grew at a fast pace, tenements squeezed in amid smaller dwellings, industries encroached on residential neighbourhoods, and skyscrapers towered above other structures. Policymakers saw a need to restrict construction size, save existing districts from devastation, and sort out conflicting activities in order to promote economy and efficiency in the city's layout and conservation of property values. The ideal traffic patterns, population densities, and levels of public infrastructure were all included in the master plans. These aims were mostly accomplished via the zoning restrictions that were first put in place in the early twentieth century. Importantly for the sake of urban development, zoning regulations separated certain uses of city space by setting maximums for building width and height and by defining permitted designs of buildings within delineated regions (zones). Consequently, the city's housing, manufacturing, and retail sectors, which had previously been well integrated, were now located in separate areas. Although zoning served to shield inhabitants from nearby harmful uses, it had the unintended consequence of making people drive longer distances to and from work and increasing the frequency with which they travelled on a daily basis, which in turn increased traffic congestion and restricted the amount of time people could spend in various parts of the city. Conflicts arose due to certain zoning regulations. Legal actions in the US contested zoning regulations that limited the availability of low-cost housing by mandating huge single-family houses on expansive lots. Exclusionary zoning was overturned by courts in many states, leading to the passage of corrective laws in others.

Subdivision regulations, which brought the first governmental regulation of unoccupied land layouts into play, evolved in tandem with zoning in the US. These rules mandated that all newly constructed roadways adhere to the master plan of the city, which in turn impacted the architectural style of new constructions. Developers were obligated to provide the necessary land for roadways, parks, and school sites as well as cover the whole or a majority of the construction costs in accordance with certain subdivision rules.

New towns

Many European nations, including the Soviet Union, Germany, the Netherlands, and France, embarked on the construction of new towns (extensive new constructions outside city centres) as state-run businesses after WWII. Governments built these new towns to contain urban sprawl inside planned projects instead of letting it expand in an unchecked fashion, since they were worried about what they saw as excessive urban density. With the exception of the Soviet Union, most of these urban areas were mostly residential suburbs; nevertheless, low-rise conurbations were able to attract both people and industry in some British cities like Milton Keynes. The Swedish government was successful in building mixed-income, accessible high-rise communities. As a low-rise ensemble in metropolitan Helsinki, Finland, Tapiola embodied many of Howard's unique concepts and included top-tier architecture. But most new town developments in Belgium, France, Italy, and Spain ended up being huge, unappealing high-rise housing complexes for the working class on the outskirts of cities.

The establishment of new towns in the United States after World War II was mostly driven by private initiative. Well-known examples include Reston, Virginia; Columbia, Maryland; Irvine, California; and Seaside, Florida. But before these endeavours, there were a handful of tiny, privately planned suburbs. One of these was Radburn, New Jersey, constructed in 1929 in accordance with ideas proposed by Clarence Stein and Henry Wright; another was Riverside, Illinois, a planned community outside of Chicago, designed by Frederick Law Olmsted in 1868–1869. Le Corbusier's Chandigarh in India, the Middle East, and South America are only a few of the remarkable instances of planned new towns in geographically dispersed regions.

Big, crowded cities sprang up all throughout Asia as a result of the region's burgeoning industrial industries in the decades after WWII. Massive construction projects including skyscraper office buildings, retail malls, luxury residences and hotels, and new airports were undertaken by several Asian countries to tackle the issues of fast development. Located across the Huangpu River from Shanghai's historic downtown centre, the Pudong New Area is a planned central business district, industrial estates, and residential areas developed by the Chinese government in little over a decade. But many emerging nations are still worried about politics and the economy, so they haven't done much to create an environmental planning function that can prevent the unsanitary conditions that plagued Western cities in the 1800s.

2 The Scope of Planning

The impact of planning grew throughout Europe in the early twentieth century as a number of national and regional regulations progressively regulated new construction. Housing for the working class became an issue for European governments, and choices on where to build homes affected the expansion of cities. Zoning as a method of municipal planning in the US originated in 1916 with a statute passed in New York City. However, federal involvement in housing and land use did not occur until the Great Depression of the 1930s. The United States and Britain developed their most comprehensive planning frameworks ever during World War II due to military mobilisation and the need to coordinate defence manufacturing. Despite its demobilisation after the conclusion of hostilities, the wartime agencies established a standard for national economic and demographic planning, although one that was much more comprehensive in Britain compared to the US.

3 Post War Approaches

After WWII, European governments launched enormous housing and reconstruction programmes in their war-torn cities. The teachings of Swiss architect Le Corbusier, art and architectural historian Siegfried Giedion, and the International school with its roots in Germany's Bauhaus were the foundation of the modernist planning principles promulgated by the Congress International d'Architecture Moderne (CIAM). The constructions constructed at this time mostly consisted of high-rise buildings with green areas in between. They were designed to be mass-produced at low cost, which was driven by the necessity to take use of new materials and technology while also satisfying the architects' need for universally replicable models. Instead of depending on regulatory tools to limit private developers, the public sector was able to exert greater direct influence over the pattern of urban expansion via its investments when the government became involved in housing building.

Specifically, Leslie Patrick Abercrombie's Greater London Plan proposed enclosing the metropolitan area in an unbroken greenbelt, constructing new towns outside the greenbelt to reduce urban population densities, and constructing circumferential highways to redirect traffic away from the central business district. Besides the rest of Britain, the Scandinavian nations, Germany, and the Netherlands also embraced the idea of a clear divide between city and country. In response to the rising demand for housing, massive suburban developments were built in the US. Although private companies oversaw and paid for construction, the federal government provided tax breaks and guaranteed mortgages to entice investors. Zoning and subdivision permission, public construction of sewage and water systems, and schools were all aspects of suburban planning that occurred at the municipal level. Unplanned expansion and urban sprawl occurred because there were no planning authorities that covered whole metropolitan areas. But in downtown areas, the federal government helped fund urban renewal agencies' land clearing efforts and the building of public housing—housing for low-income individuals that is owned by the public. Only public buildings including schools, police stations, and leisure centres were allowed to undergo local government-led repair efforts. It only indicated what would be desired and depended on private investors for much of the new development. Because to the lack of interest from the private sector, several previously cleared lots remained undeveloped for many years.

4 Planning and Government

The role of city planning in municipal administration evolved in unique ways throughout nations. Nowadays, private developers often need government approval before they can break ground on a project. But in the US, they may construct "as of right" provided their designs are in accordance with the zoning regulations of the town. City planning in Europe is the purview of an executive department with considerable power because of the continent's highly centralised municipal governance. While a planning department provides executive and advising services in the UK, the elected local council is the actual authority when it comes to planning. Those who would want to develop but have their plans rejected have the option to appeal to the federal government.

In most American cities, the mayor and council have the last say when it comes to major decisions, but for day-to-day planning, an appointed commission often steps in. Private development proposals are the main focus of planning activities, which mostly include giving or negative feedback. The commission has workers that report to it in bigger cities. From 1949 to 1974, most American cities had strong, semi-independent urban renewal agencies that oversaw rebuilding plans as part of a massive, federally funded urban renewal programme. Though they are still in existence in some form, these entities often became subservient to the mayor or merged with economic development agencies, which are frequently quasi-autonomous businesses. These agencies report to an independent board of directors, often comprised of members from the business community, but they are nominated by the mayor and council. Agencies were given the green light to sign development agreements with private investors, which came in handy when the local administration started to focus on economic development planning.

India: Planning and architecture

The federal government has included city planning in its strategy to address development and social welfare in a number of nations, especially in northern Europe. Even though metropolitan governments in the US never really took off, the federal government nevertheless became involved in local planning issues in 1965 when it created the Department of Housing and Urban Development to oversee the programme and passed national housing and urban renewal laws. The emerging country governments that emerged in the 1960s and 1970s as a result of developing nations' declarations of independence from colonial powers tended to establish the ground rules for urban planning.

5 Competing Models

Several influential ideas of urban planning emerged in the twentieth century, each of which had an impact on how cities looked and felt for residents for as long as it lasted. In the middle of the twentieth century, comprehensiveness was the main objective of city planning. Land use, transit, and housing all needed to be planned in relation to one another as the interconnectedness of different parts of the city became more apparent. Academic planners were influenced by developments in management science and operations research to develop a universal method, sometimes called "the rational model," wherein specialists would assess options in relation to predetermined objectives and select the best one. The rational model had a short period of dominance until its proponents began to voice concerns about the intangible human costs of public policy choices and the scientific method used to make them.

There was a lot of backlash against the modernist paradigm, which advocated for the complete gutting of existing structures and their replacement by experts in urban planning who were cut off from public input. Jane Jacobs, an urbanologist, was instrumental in discrediting the modernist school of thought. Sarcastically referencing Le Corbusier's "towers in the park" (from his cité radieuse concept) and Ebenezer Howard's antiurban garden city, she referred to revitalised downtowns and housing projects as the "radiant garden city" in 1961's The Death and Life of Great American Cities. Jacobs said that massive cleaning efforts imposed an inhumane orderliness on communities and destroyed their intricate social fabric. She didn't consider dense population as a problem, but rather as a key component of thriving cities. She advocated for a variety of uses and demographic groupings as a core principle in regulating urban development because she thought that cities were more appealing with vibrant street life. In contrast to homogeneous urban environments, which, as Jacobs points out, rely on unsustainable exploitation—as shown, for example, in the case of timber or mining towns that fall once their rich resources have been extracted—diversified urban settings contribute to sustainable development. There were others who shared Jacobs's criticism. Opposition to the displacements generated by large-scale modernist planning began in the 1960s with urban social movements, which at times amounted to rebellion. There was strong resistance to attempts to destroy inhabited homes in cities throughout Europe and the US. Government efforts to dismantle squatter colonies in emerging nations sparked comparable counteroffensives.

Planning orthodoxy in the US and Europe started to consider Jacobs's points towards the century's conclusion. The revitalization of older structures, protection of historical sites, adaptive repurposing of old buildings, mixed-use development, and the concept of the "24 h city" (i.e., areas where many purposes would generate constant activity) were all given more priority. In recent years, there has been a noticeable trend towards constructing large-scale new developments on formerly inhabited or "brownfields" properties, such as abandoned industry districts, obsolete port facilities, or railway yards. The modernist ideas of the past, nevertheless, were still very much alive and well in emerging nations. So, for instance, China displaced a large portion of its urban population to build roads and sports facilities for the 2008 Beijing Olympics, and it also created new business districts by constructing high-rise buildings following the functionalist Corbusian paradigm.

6 Contemporary Planning

At the turn of the millennium, planning practices did not adhere to any one model of a repeatable procedure or an ideal result. Some American and European towns did what was asked of them and adopted a participatory method of planning that included the people most impacted by proposed changes. The idea of participatory planning has gone global, but its implementation is still in its infancy. In most places, the level of democracy is directly proportional to the level of public involvement in planning. Planning follows an autocratic regime. In a more collaborative model, the planner's function shifts from expert to mediator among various interest groups, or "stakeholders." Proponents of the idea of "communicative rationality" have lent their support to this recast function. But some who disagree with this idea say that it might lead to results that aren't in the public's best interest since it could stifle innovation or only benefit the powerful. Additionally, they are worried that if neighbours can say "not in my backyard" ("NIMBYism") whenever they feel their property values will be negatively affected by development, it would be impossible to provide cheap housing and necessary public infrastructure.

There is a considerable deal of diversity in modern urban planning due to the planners' enormously diverse project kinds, the absence of agreement over procedures and objectives, and the diverse methodologies used in various cities and nations. Even while many localities still adhere to the basic notion of rigorous use segregation, there is a clear tendency towards mixed-use development in urban centres, especially of complementing activities like housing, retail, and entertainment.

7 Changing Objectives

Focuses among the several aims of planning have shifted, but some, like environmental preservation, continue to be vital. Economic development planning has been more prominent, particularly in older communities hit hard by the industrial sector's fall. Economic development planners are a lot like marketing executives: they tout their cities to investors and judge physical development based on how appealing it is to capital and how many jobs it can create, not on how well it follows a master plan or how healthy it is. Developers and businesses that support local economies are the focus of these planners as they seek to get development agreements. In many countries, like the US and UK, planning authorities are actively engaged in making arrangements with private developers in an effort to promote economic growth. To provide one example, in the UK, developers may be able to construct on a piecemeal basis in exchange for community amenities, money, or other advantages, or "planning gain" from the government. Public subsidies are often included in U.S. negotiations, even if special authorization is not necessary provided the structure falls inside the zoning code. In a typical development agreement, the government promises to invest in the region or provide amenities in exchange for land, tax breaks, or leniency from regulations. Another possibility is for the city to reach an agreement with a private company that would allow the company to enter or stay in the region in exchange for certain privileges. Controversy surrounds many of these deals, particularly when a city uses its eminent domain power to acquire privately held property for development purposes.

People are interested in a planning movement from the late twentieth century that offers different perspectives on suburban development; this movement goes by many names: new urbanism, smart growth, and neo-traditionalism. This movement has supported new construction that brings home, work, and shopping closer together, encourages pedestrian traffic, promotes development around mass-transit nodes, and mixes types of housing. It reflects a strong dislike for urban sprawl, suburban traffic congestion, and long commuting times. By endorsing the new town of Poundbury in Dorset, which has a traditional aspect, Prince Charles became a staunch advocate of neotraditional development in the UK. In a country like the US, where urban expansion and dispersion were constant problems, such initiatives mostly emerged as little pockets of planned development. While the movement has mostly had an impact on new suburban construction, it has also been used to revitalise existing districts in the US and UK. A number of HOPE VI schemes in the US, which entailed the demolition of public housing and the construction of mixed-income developments, as well as Paternoster Square in London, which is next to Saint Paul's Cathedral, were built in line with neo-traditional or new urbanize principles.

8 New Diversity

Several phenomena have contributed to the gradual disintegration of once-universal rules governing adequate planning. To begin, there has been a rise in the prominence of academic arguments that reject the idea of a "one plan fits all" strategy. A more democratic approach that is more attuned to local heterogeneity has supplanted the earlier consensus on the idealised modernist model of orderly development that embodies use separation and standardised building. The second point is that many areas just do not have the means to implement standards, even if doing so would be a good idea in certain cases. Despite planners' earlier condemnation, informal marketplaces and settlements in developing countries seem inevitable and, in many cases, suitable for meeting the requirements of low-income populations. As a result of pressure from foreign assistance organisations, planners in these areas are trying to improve existing slums and street markets instead of eradicating them. Third, planners have been compelled to seek market-based solutions to issues like pollution and the provision of public services by political forces that advocate for the free market. Instead of relying only on regulations, this has prompted the privatisation of utilities and services that were formerly held by the public, as well as the trade of rights to develop property and release pollutants.

A city that is well-designed and extensively planned has the potential to improve society, which was an underlying assumption of planning from its inception. The bias was towards environmental determinism, to rephrase. Subsequently, planning has taken a more measured approach, and the idea that physical surroundings may significantly impact social conduct has faded. Nevertheless, public policy is a tool that planning depends on to create a more equal and appealing environment. This approach does not drastically change human conduct, but it does contribute to changes in the quality of life for many people.

Sustainable Development Goals (SDGs)

Biman Gati Gupta

Governments and associated public bodies need to pay attention to and concentrate on a lot of topics and concerns in society. Given the recent acceleration of both industrialization and urbanisation, the need of prioritising sustainable development cannot be overstated. One definition of sustainable development is progress that does not compromise or harm existing social norms and values. To rephrase, it has been seen as taking care of current demands at the expense of future generations' capacity to do the same [1]. Poverty, environmental degradation, climate change, inequality, injustice, and many other difficulties have been faced by societies across the world. More sustainability and a brighter future for everyone might be achieved by following the SDGs in this area. A thorough evaluation of the local community and environment is necessary for the creation of more effective SDGs. Sustainable development strategies may need to adapt to the specific challenges faced by each nation or city [2].

Sustainable and equitable health for everybody at every level is one of the SDGs. Reduced poverty, a pristine earth, and happy, peaceful people are all part of sustainable development's larger goal and strategy. While addressing current needs, sustainable development also pays close attention to what may happen in the future. The term "sustainable development" refers to both a method and an idea that encourages corporate and individual actions that benefit communities and society as a whole [3]. Considering at the larger context, a government that adopts a sustainable development strategy has a better chance of ensuring the safety of its citizens and the nation as a whole. Aiming towards sustainable development necessitates prioritising social progress alongside economic growth and environmental balance. Education, fundraising, volunteering, empowering change-makers, and many other initiatives are crucial to achieving the SDGs [4].

B. G. Gupta (✉)
Elitte College of Engineering, MAKA University of Technology Sodepur, Kolkata 700113, India
e-mail: bimangupta@yahoo.in

B. G. Gupta (ed.), *Urban Development with Treated Recycled and Reuse of Wastewater: Sustainable Management*, Urban Sustainability,
https://doi.org/10.1007/978-981-96-0857-7_4

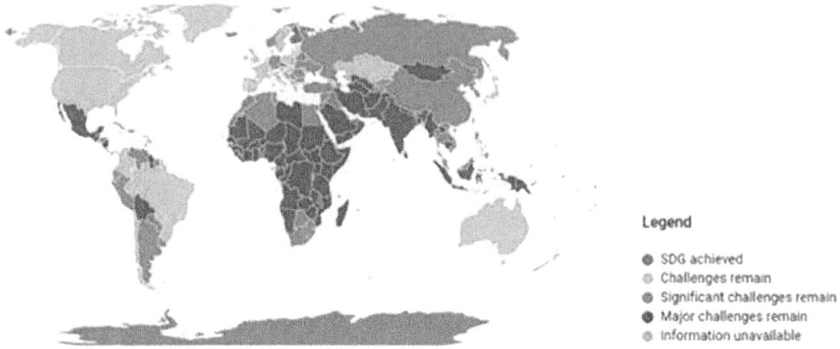

Fig. 1 Status of SDG 6 (clean water and sanitation), from https://dashboards.sdgindex.org/

In 2015, the UN General Assembly adopted the Sustainable Development Goals. By establishing 17 interdependent goals, a better future for everybody was advocated. Some of the 17 interconnected goals include ending poverty, improving health, ending hunger, promoting gender equality, improving education, growing the economy, creating sustainable communities, combating climate change, making responsible consumption and production choices, fostering peace, establishing industry, fostering innovation, and improving infrastructure [5]. Incorporated within the 2030 Agenda, a United Nations Resolution, are the SDGs. In order to improve the associated consequences and results, these particular goals are being thoughtfully evaluated and are expected to be completed by 2030. The particular resolution lays out the goals for each target and the measures to be used for measuring progress towards each aim. While some of the Sustainable Development Goals (SDGs) are intended to remain in place beyond 2030, the bulk of the SDGs are anticipated to be completed within that time frame. With the various supplied tools and methods, it may be easier to monitor the targets [6].

According to the most recent data from the UN SDGs dashboard, the majority of states continue to encounter significant difficulties in achieving the SDGs targets. As seen in Fig. 1, the only three countries achieving SDG 6 successfully are Croatia, the Czech Republic, and Finland. However, when looking at the details, it becomes clear that these three countries are not meeting the objectives set for SDG 6 [7].

1 Wastewater Contribution into SDGs

When it comes to development, water is one of the most important and crucial resources. Water is becoming more limited and contaminated as a result of growing demand brought about by urbanisation, industry, and population development. Changes in rainfall patterns and other characteristics of the climate as a result of naturally occurring human interference have the potential to amplify the effect. The world's water crisis has escalated into a serious problem, impacting people's lives and the environment in unpredictable ways [2]. Not long ago, wastewater wasn't thought of as having much value; nevertheless, nowadays, it is seen as a potential new supply of potable and non-potable water. Filtration and treatment of wastewater allows for its reuse for human consumption and a wide variety of other functions, including irrigation, cleaning, industrial processes, and more. There are a lot of current and future uses for recycled wastewater, and it might give many different kinds of advantages to society [8].

There is still room for improvement in the current understanding of wastewater's potential uses; further investigation into these areas might greatly increase the advantages already obtained. Society has benefited greatly from wastewater conversion into reused water on several levels. Extensive research into potential uses for wastewater has engaged a wide range of interested parties [9]. There are a lot of ways wastewater might be put to use in the context of sustainable development goals, such improving cleanliness and making it drinkable. We need to make progress in wastewater management to enhance the related results since poorly managed wastewater might have many negative effects on society. Among the seventeen Sustainable Development Goals (SDGs), SDG 6—"ensuring availability and sustainable management of water and sanitation for all"—has direct bearing on wastewater treatment. Through attaining this goal or even attempting to partially achieve, it would lead immensely benefit people on a global level. The significance of clean water for improved quality of life and socio-economic development, inculcating environmental protection and health, is considerably high. Having said this, attempting to make use of wastewater and bringing it to optimum utilisation could be regarded as highly critical [10]. The sixth goal of sustainable development, which is related to clean water and sanitation focuses on improving water quality by reducing the amount of wastewater that is not treated and enhancing safe reuse and recycling globally. Through this, there will be more clean water available for use for individuals which will also lead to huge progress on sanitation and wastewater management. Considering the scarcity of this resource, clean water would never be available if wastewater is not gathered, treated, reused and disposed of adequately. Regardless of its size, every country today is in need of reusing wastewater and putting it to the best possible use by all means [11]. The contribution of wastewater treatment to SDGs can be summarised in the Table 1. The wastewater treatment plant is essential to achieve eleven SDGs from the 17 SDGs. The main contribution of the wastewater treatment plant is coming from its capabilities on providing clean water and increasing water availability.

Table 1 Contribution of wastewater treatment into SDGs

Related SDG	Contribution
SDG 1: no poverty	• Increasing the income of the smallholder
SDG 2: zero hunger	• Increasing the agricultural productivity by reducing the clean water price and recycling water for agricultural purposes
SDG 3: good health and well-being	•Reducing potential diseases • Reducing bad odors
SDG 6: clean water and sanitation	• Increasing the water availability • Improving the water quality • Discharging wastewater safely
SDG 7: affordable and clean energy	• Increasing the energy availability by providing clean water • Production of biogas from wastewater treatment plants
SDG 8: decent work and economic growth	• Increasing the gross domestic product (GDP) by reducing the cost of water which is essential for all production • Recycling industrial wastewater decreases the operational cost • Recycling valuable products from industrial wastewater streams
SDG 9: industry, innovation and infrastructure	• Providing clean water for a small-scale business in the rural area • Reducing materials ecological footprint • Adding value to the wastes
SDG 11: Sustainable Cities and Communities	• Improving the waste management system • Reducing urban environmental impacts by circulating the water • Improving the air quality by reducing the bad odor arises from lack or inadequate wastewater treatment process
SDG 12: responsible consumption and production	• Improving the waste circulating process • Recycling of salts and valuable materials/ products from the effluent wastewater
SDG 13: climate action	• Reducing the GHG emissions • Production of biogas
SDG 14: life below water	• The efficient wastewater discharge minimizes the negative impacts on the marine life

2 Proposed Indicators to Improve the Wastewater Facility into the SDGs

When it comes to the SDGs, the wastewater treatment plant is a mixed bag. Although most heavy metal levels in treated water from sewage treatment plants were within acceptable limits, some studies found elevated levels of Cd, Ni, Pb, and Cr in the treated water and pesticides like BHC and DDT in the soil that received the wastewater for irrigation. This is likely a result of the long-term use of treated water for irrigation. The livestock and humans both eat the grains and veggies. So, any waste management strategy has to prioritise watering with treated effluent immediately. In addition, soil deposition occurs when wastewater with a high bacterial biomass or solid loading is used [12]. "Inactivation decreased the stability of THM4 and HAA9, but chlorate remained constant following the inactivation process, according to another evaluation of the hazards of disinfection byproducts, primarily trihalomethanes (THMs), haloacetic acids (HAAs), and chlorate discharged from wastewater treatment plants (WWTPs) using either single or combined disinfection procedures. Disinfection by-products were entirely reduced when WWTPs switched to peracetic acid or peracetic acid/Ultraviolet as disinfectants, while chlorine alone or in combination with UV significantly raised levels of HAAs, THMs, and chlorate in the treated water [13]. The treated water from WWTPs may be less harmful if it is treated properly and not used for an extended period of time.

Nevertheless, a collection of indications was suggested to mitigate the unfavourable effect; they are shown in Table 2. Literature reviews [14, 15, 14–18] formed the basis for the development of the suggested indicators.

3 Wastewater Relation with Each SDG

Clean water and sanitation for everyone by 2030 is the goal of Sustainable Development Goal 6, which was created with this purpose in mind. The sections that follow provide a detailed discussion of the relationship between wastewater treatment and other Sustainable Development Goals Targets. The connection between SDG 6 and the other SDGs is seen in Fig. 2. The details of the SDG interlinking tool are available in [19].

3.1 SDG 1: No Poverty

Poverty and water scarcity go hand in hand since water is essential for so many things—growing food, maintaining good health, working, constructing a home, and even attending school. A lack of water makes it more difficult, if not impossible, to escape poverty. People and society rely on water for a wide range of purposes. Almost every establishment in society, including homes, businesses, and educational institutions, needs water to function [20]. Furthermore, communities could fail to steer themselves towards prosperity and advancement. More people will fall into poverty as a result of the decreased operating effectiveness of these institutions. It is essential for a nation or civilization to operate well and have good working conditions overall if it wants to advance. Since water is essential for growing food, it follows that farming and agriculture also need it. There will be deterioration in living

Table 2 The proposed indicators in wastewater treatment facility to increase contribution into SDGs

SDGs	Indicators
Goal 1. No poverty	1. Total Taxes paid 2. The number of employees 3. Average employees' salary 4. Number of training programs per employee
Goal 2. Zero hunger	5. Nature of land occupied 6. Disaster risk management policy 7. Total nutrients, organic matter, and carbon recycling
Goal 3. Good health and wellbeing	8. The number of accidents 9. Dust control measures 10. Total healthcare benefits given to the employees 11. Calculating human toxicity potential, particulate matter, and photochemical ozone creation potential
Goal 4. Quality education	12. Availability of assessment programs to test employees' skill levels 13. The average number of training programs conducted per employee 14. Abolition of all child labor
Goal 5. Gender equality	15. Percentage of women recruited and employed 16. Income paid to women 17. Percentage of leadership positions occupied by women
Goal 6. Clean water and sanitation	18. Water pollution and water reduction, wastewater management, and circular water strategies 19. Water consumption 20. Nonportable water consumption 21. The total reduction in water consumption 22. Decrease the water footprint 23. Quantifying freshwater aquatic ecotoxicity potential and eutrophication potential
Goal 7. Affordable and clean energy	24. Conducted an energy balance analysis 25. Degree of energy sharing
Goal 8. Decent work and economic growth	26. The investment in clean technology research and development 27. Pay of goods from local communities 28. The proportion of housing offered to temporary employees 29. Employment creation

(continued)

Table 2 (continued)

SDGs	Indicators
Goal 9. Industry, innovation, and infrastructure	30. Conducted social, economic, and environmental impact assessments 31. Applying circular business concept 32. Tracking and reporting climate impact 33. Overall value-added
Goal 10. Reduced inequalities	34. The proportion of training programs for marginalized populations 35. Total wage disparities between different working groups 36. Level of diversity and inclusion
Goal 11. Sustainable cities and communities	37. Resource's sustainability strategies 38. Participation in microgrids 39. Overall produced waste
Goal 12. Responsible consumption and production	40. Tracing all types of short-term and long-term pollution, such as GHG, methane, and water pollution 41. Overall resources and materials required 42. Overall waste treated 43. Materials (phosphorus, nitrogen) recycling potential
Goal 13. Climate action	44. Reporting all the types of pollution 45. Location of the facility
Goal 14. Life below water	46. Tracking and reporting all water-related pollution 47. Quantifying the marine ecotoxicity
Goal 15. Life on land	48. Quantifying biodiversity the impact 49. Effective methane yield
Goal 16. Peace, justice, and strong institution	50. Anti-bribery management systems 51. Stakeholders' engagement 52. Level of compliance to projects' directives and provide transparent information 53. Sum of environmental violations
Goal 17. Partnerships for the goals	54. SDGs' incorporation in business strategies 55. Collaboration with the different agencies

standards as a result of farmers' inability to cultivate crops and other associated activities due to a lack of water. The availability of water is crucial to a society's growth since it allows for the improvement of infrastructure [21]. On a general level, water is an essential ingredient for almost every human activity. Maintaining access to clean water is critical for every civilization that aspires to develop and advance. As an example of a notion that might help nations and communities satisfy the water availability needs of connected jobs, wastewater reclamation has been highlighted. As wastewater reclamation and reuse technologies advance, those with a stake in the matter are taking part in procedures that might put water to good use in

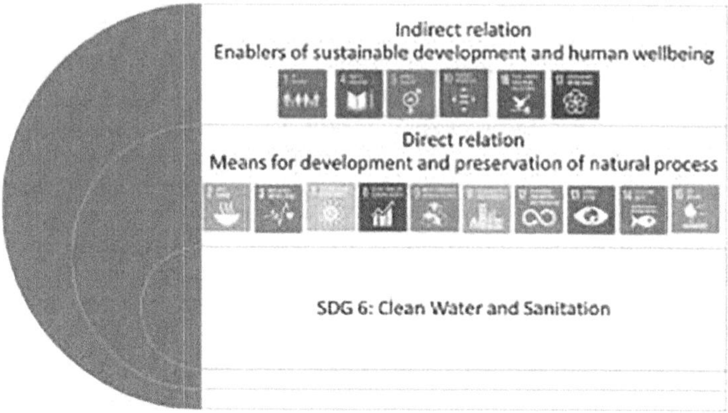

Fig. 2 The relationship of SDG 6 with other SDGs

a variety of contexts [22]. The hiring procedures of wastewater treatment facilities provide employment possibilities, which in turn reduces the unemployment rate [23]. Urea fuel cells [24–27], bio-electrochemical systems [28, 29], energy "electricity" generation [30, 31], or chemical synthesis [32, 33] are only a few examples of the many recent advancements in wastewater treatment. By reducing the price of fresh water, these solutions will eventually benefit society's economy. For smallholder farmers in particular, the wastewater treatment facility's ability to produce low-cost, high-quality fertiliser is a major plus [34, 35]. As a result, their revenue and net profit will rise [36–38].

3.2 SDG 2: Zero Hunger

The years 2015–2030 are currently considered pivotal and significant for global sustainable development to achieve progress. In order to build a robust food system and provide nutritional meals, access to clean water is vital. It is essential to proper nutrition to ensure that everyone has access to clean drinking water. The health of many individuals throughout the globe is jeopardised because they cannot get clean water to drink. To keep to a healthy diet, it is essential that people have access to a sufficient amount of clean water to drink every day [39]. There are several important aspects, including food, hunger, and nutrition, but irrigation is also crucial. It has long been known that wastewater may provide a significant amount of water for agricultural purposes. Nutrients and water are well-known to be complementary.

It is critical for food production to have access to enough water for agricultural operations and procedures. Most individuals cannot grow enough food to satisfy their requirements due to water scarcity or stagnation. Given its ability to filter and recycle water for use in agricultural irrigation, wastewater might play a significant role in this context. The number 143. In addition to providing for their families' needs, the impoverished would benefit from having access to sufficient water for agricultural purposes [40]. Wastewater recovery has the potential to provide enough water to meet human needs, including dietary needs. The increasing need for drinkable water in agricultural settings might be addressed by wastewater treatment. Wastewater reuse and recycling has improved greatly thanks to recent technical developments, allowing for more effective and efficient attainment of the intended results [39].

The competent authorities must pay close attention to undernourishment, which is one of the other worldwide prevalent challenges. The problem of undernourishment is especially prevalent among those living in slums and other underdeveloped regions [41]. Chronic undernourishment, which occurs when people do not have enough food to meet their nutritional needs, is similar to hunger. Individuals experience more hunger as a result of reduced food growth and production caused by low water availability in different regions. Wastewater has the potential to enhance and support crop development if it is properly treated and used via recycling and reuse. Assuming this takes place, solving the problem of hunger and reaching the goal of zero hunger will be much easier [42].

In addition, the treated wastewater improves agricultural yields in the regions that receive it due to the elevated concentrations of potassium (K), nitrogen (N), and phosphorus (P). Approximate nitrogen production from one metric tonne of generated sludge was 14 kg, phosphorus production 6.75 kg, and potassium 4.25 kg [12]. Food security is under risk due to diminishing supplies of phosphorus-based fertilisers, which is why there is a growing interest in recovering this crucial macronutrient from wastewater [43].

3.3 SDG 3: Good Health and Wellbeing

Parasites, fungus, bacteria, and viruses in raw wastewater may cause illnesses in the intestines, lungs, and other parts of the body [44]. Additionally, the germs that cause it might cause fever, diarrhoea, cramps, and headaches. Wastewater has the potential to create a host of health problems; hence, there is an urgent need to find a solution. [45, 46]. The chance of illnesses is greatly reduced with the use of contemporary WWTPs. The wastewater treatment procedure gets rid of disease-causing bacteria and other dangerous organisms [47]. People have been dealing with a lot of problems due to the fact that little has been done on this aspect, even though these dangers and possible adversities are associated with wastewater [48]. In addition, reducing wastewater odours is a key advantage of WWTPs. The result of wastewater treatment plants is an increase in odours; however, with the widespread use of nanotechnology and other modern technologies, this impact has been much mitigated in recent years. [36, 47, 49, 50].

The amount of sewage has grown in tandem with the population, and this has resulted in the introduction of more harmful chemicals and compounds that impact human health. These treatments are crucial for human health since sewage and wastewater treatment problems have been on the increase and killed a lot of people throughout the years. Due to improperly managed or treated wastewater, some individuals have been sick or even perished [51]. The health hazards linked to wastewater have had a more profound impact on those living in rural areas compared to those living in urban centres. In rural regions, there is a greater prevalence of illnesses transferred by wastewater due to the poor levels of management and treatment.

Conversely, urban regions have comparatively better water management and have managed their challenges to a lesser degree. Wastewater treatment has the potential to improve people's health and wellbeing, given the correlation between the two. If this area is improved, people all around the world may be able to keep their health and prevent diseases that are associated with it [40].

3.4 SDG 4: Quality Education

The situation is becoming worse by the day, and there is enough of fresh water on Earth, but people can't seem to get their hands on it. The water supply is unregulated and uncontrolled, and the economy is in shambles. The majority of developing world residents still reside in rural regions, and many rely on agriculture for a living. Higher education and other occupational skills may assist these individuals overcome these problems; we must secure them access to them if we wish to accomplish the SDGs. The wastewater issue is limiting the water supply to farmers and increasing the contamination of the fresh water supply, according to Cooper et al [52]. As a result, agricultural goods are either destroyed or have their quality diminished. These factors have a negative impact on rural families' income, which in turn causes poverty. As a result, child labour becomes more common as youngsters, whose families cannot afford to send them to school, are compelled to work in cities instead.

3.5 SDG 5: Gender Equality

Every aspect of a man's life is affected by water, as it is an essential human need. Contrarily, women face greater obstacles as a result of wastewater rules than men do. Without proper wastewater treatment, women are disproportionately impacted. Since the majority of women's jobs include domestic duties, they are particularly vulnerable to exposure to hazardous wastewater in the course of their daily tasks, whether at home or on farms. Most women do unpaid or poorly compensated labour in numerous fields, including farm and household labour, which is a painful social reality. Herrera [53] claims that people in this demographic are more likely to drink unclean water because of their poor income.

3.6 SDG 6: Clean Water and Sanitation

In order to maximise the supply of pure water, wastewater treatment is crucial. Since wastewater and sewage may cause a host of health problems, access to clean water has become an absolute need in modern society. Clean water is essential for communities and overall development prospects, apart from individual hardships. Water might be seen as a crucial component in societal evolution. It is now more important than ever to guarantee that there is an abundance of potable water since, without it, development and progress are almost impossible. Wastewater treatment plants are now vital in modern sanitation and disease control because they remove parasites, viruses, and dangerous bacteria [48]. There has been a lot of interest and activity among governments and society in creating technologies that might recycle water and make it clean once. Wastewater management relies on clean water and sanitation, thus efforts are focused on these areas to maximise their advantages. Effective wastewater management is now absolutely necessary due to the very significant dangers connected with wastewater [54].

The society's environment is affected by the poor and insufficient treatment facilities, which are present in many different places [55]. When it comes to sewage and sanitation, the government must step in. The government's policies and programmes may have an effect on wastewater management, which in turn may have an effect on society as a whole. Sanitation is the procedure that reduces the hazards and damages that people may experience as a result of wastewater, so the two might be seen as having a close relationship. Effective wastewater treatment is critical for the availability of potable water. A clean and risk-free method of minimising any damage to groundwater quality must be guaranteed [56]. People in the community may reap the benefits of more readily available clean water if efforts are made to improve this situation [57].

3.7 SDG 7: Affordable and Clean Energy

To get to efficient energy use in water treatment, many places throughout the globe need new regulations and investments (Fig. 3). The availability of potable water is crucial to the production of hydropower, which is regarded as one of the world's top power producers.

Renewable energy use across the world's sectors rose 1.5% in the first quarter of 2020 compared to the same period in 2019 [58], according to reports. Global utilisation of renewable energy is projected to rise by around 1% in 2020, according to the projections. Many organic compounds found in wastewater have potential as renewable energy sources. It is estimated that the internal energy of wastewater is around 6.3–7.6 kJ/L (1.75–2.11 kWh/m^3) [59]. Hydrothermal treatment of wastewater sludge is a typical practice for reducing complex organic molecules to simpler ones, which the anaerobic digester may then employ to produce biogas. After

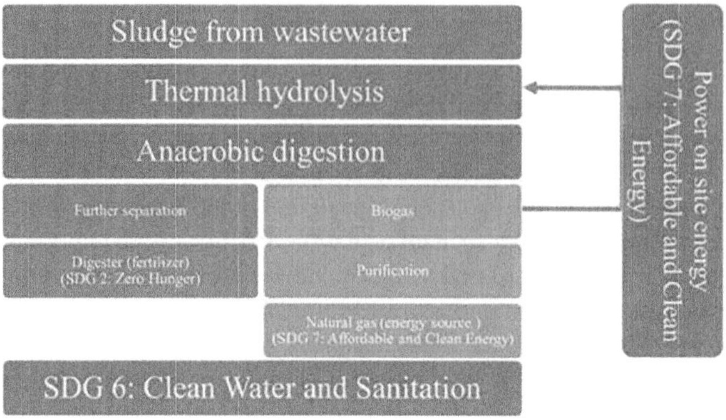

Fig. 3 Wastewater to *energy system* overview

preliminary treatment, the biogas may be used in combustion processes directly or in fuel cells to generate power [60, 61]. An other strategy for enhancing the biogas's energy density is methanation [62]. The biogas is converted into an organic fertiliser of superior quality, which is then utilised in farming. The suggested methods are summarised in Fig. 3 [63]. Since utilities are able to supply more economical energy to the communities thanks to the utilisation of wastewater resources, the price of power production has dropped. A case study of a power plant that saved around US$18 million over six years was the San Luis Potosi facility in Mexico. Instead of using groundwater, which increases prices by 33%, treated wastewater was used instead. In addition, the annual cost savings for transporting biosolids at the Cusco wastewater treatment facility are around $230,000 [64].

3.8 SDG 8: Decent Work and Economic Growth

Inadequate investment and a lack of development initiatives are key obstacles to acceptable employment prospects worldwide, particularly in impoverished and emerging nations. The leaders of any nation must devise plans for long-term economic expansion that provide their people with good employment opportunities, which in turn boost economic development and propel the nation forward. Wastewater management is very important for everyone's sustainable growth. Increasing the number of treatment plants and systems in the healthcare and other water-dependent industries, as well as attracting new investors to the wastewater treatment plant, may provide long-term job prospects. As a result, many new occupations will emerge, each calling for a unique set of abilities, expertise, and background. The government can analyse job quality using various data sets pertaining to wastewater and employment. This will help them enhance job quality in the wastewater sector and provide people with decent work opportunities that prioritise their safety, health, and profile. Improving wastewater treatment facilities worldwide would most certainly lead to a large increase in the labour force, which in turn would attract a wide range of skilled workers [65]. Investors funding the treatment facilities will encourage the operation of several companies that provide the equipment needed by these plants, which is good for the country's economy in many ways (Fig. 4). Developing nations in particular will see a reduction or elimination of the unemployment rate as a result of the country's economy receiving more foreign investment. A rise in GDP is the end result of this. One country where this is evident is China, where data from 30 provinces reveals a correlation between effective wastewater treatment and GDP [66]. Moreover, because most industrial facility depends on water within their operation lowering the cost of the water or recycling will lead to reducing the overall operation cost [67]. The wastewater treatment plant may also reduce the chemical oxygen demand (COD) of process wastewater by recycling items like crude proteins and other value-added goods that can be extracted from poultry effluent using ultrafiltration [68]. The progress towards the release of valuable biomass from wastewater treated with photosynthetic bacteria (PSB) is encouraging. This PSB possesses the ability to create additional valuable materials like carotenoids, bacteriochlorin, Coenzyme Q10, polysaccharides, polyhydroxyalkanoates, 5-aminolevulinic acid, polyhydroxy chain alkyl esters, bacteriocin, and single cell proteins. These materials could find use in various industries, including chemical, agricultural, medical, food, and wastewater treatment [69]. In order to make FeCr2O4, the Cr(VI) may be extracted from wastewater. The calcined ash, which includes Cr_2O_3 (45 wt.%), can be reused instead of chromite ore [70].

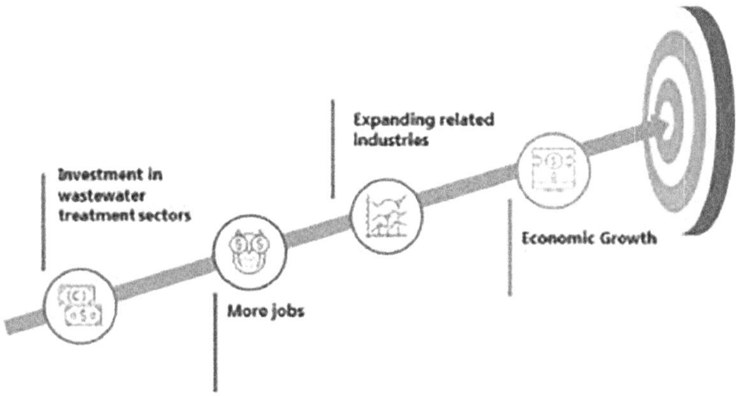

Fig. 4 Contribution of wastewater treatment to the SDG 8: decent work and economic growth

3.9 SDG 9: Industry, Innovation, and Infrastructure

Industries, infrastructure, and the encouragement of innovation across many technologies determine a nation's growth and advancement, particularly economically. Wastewater treatment is one area where innovation, infrastructure, and industry are bringing about significant strides forward. When it came to developing new wastewater technologies, many working groups put money into making various instruments and procedures. Aside from that, there are a number of state-of-the-art developments happening in wastewater technology, such as microbial fuel cells [71–73] that generate both wastewater and electricity, biohydrogen and wastewater treatment [74–76], wastewater treatment and water desalination [77], and wastewater treatment and chemicals production [32, 78]. These developments provide hope for new approaches to the challenges of wastewater treatment cost management and profit maximisation. This Sustainable Development Goal seeks to facilitate technical advancement so that different nations may address their economic and environmental problems. Qu et al. [79] states that wastewater is the primary cause of environmental impact. Thus, we must now develop companies, engage in scientific research and innovation, and provide an appropriate infrastructure and technology to reduce the environmental threats caused by wastewater. This will have several positive effects, including a cleaner environment and less environmental impact, more job possibilities, and cheaper water for small companies in rural areas [80, 81]. In the long run, this contributes to the country's economic growth and development. Wastewater treatment also has the added advantage of increasing the waste's monetary worth. Treatment wastewater treatment will add value to the supply chain of the olive oil sector in Greece, according to their study [82]. Similar studies were conducted for oil palm and other products [49, 67].

3.9.1 SDG 10: Reduced Inequality

One of the most pressing issues facing the contemporary world is inequality, both between nations and among people. One of the primary goals of the SDGs is to eliminate disparities by guaranteeing that no one is disadvantaged. Minimising trade disparities and edges, lowering relative income, and other forms of international inequality are all interrelated. The goal of leaving no one behind, as stated by Street et al. [83], is to ensure that those with lower incomes have the chance to take part in cross-border commerce, which in turn helps the economy thrive by attracting investment. In spite of the United Nations' best efforts, global inequality continues to worsen. Several nations are experiencing a severe water shortage, and the situation is particularly dire in the world's least developed nations due to their inadequate infrastructure and lack of access to modern technologies. Inadequate infrastructure for wastewater treatment and management exacerbates the problem and costs many nations money and harms the environment. When comparing industrialised and developing countries, there is a huge disparity in access to the equipment and technology used in the wastewater treatment facility. As a result, individuals in developing nations feel the brunt of health and environmental problems that manifest regionally differently. Research has shown that impoverished nations have a much higher risk of child mortality from water-related illnesses compared to industrialised nations [84]. Wastewater treatment is crucial, but the most developing nations cannot afford the costly equipment and methods needed to do so. Because of this, the impoverished and other vulnerable populations in these countries are more vulnerable to water-related illnesses. The installation of wastewater treatment facilities is a challenge for most developing nations since investors in this field are wary of putting money into countries with unstable economies. Thus, the few employment prospects in these nations have a devastating impact on the standard of living of the average citizen and the already precarious state of the national economy. Because of this, there is a significant disparity in terms of national equality.

3.9.2 SDG 11: Sustainable Cities and Communities

A sustainable development in infrastructure, economics, and all other areas of cities was necessary to accommodate the fast expansion of cities caused by increasing population and migration towards cities. El Zein et al. [85] states that when a community achieves sustainability, it allows and promotes chances for business for its residents. Additionally, it offers people a place to live that is both secure and economical, which boosts the city's reputation in terms of commerce and other economic indicators. Different activities in different industries will be enhanced by an increase in population. To meet the growing demand for houses, for instance, the building sector will ramp up its operations, among other things. The resources and raw materials necessary to carry out these operations are essential. Along with home usage, water is a vital resource used in all these activities. The situation is becoming more dire in urban areas as the worldwide demand for water continues to rise, according to a number of publications [1]. Water management in urban areas is a top priority for municipal administrators. Wastewater treatment and management is essential for urban areas to keep their climate and environment sustainable. It opens up a world of possibilities for wastewater treatment since, when used correctly, it can be turned into a cheap and useful source of water, energy, and other consumables. Since cities are the primary locations for wastewater treatment, dealing with this issue may be rather difficult (Cornejo et al., 190). The city administration has a formidable challenge in controlling and treating the massive volumes of wastewater generated by both industrial and residential sources. Thus, it is crucial for city officials to design the fabrics and infrastructure of the city in a manner that can handle the problems associated with sewage treatment and administration. However, most cities' infrastructure would be crippled and people's lives will be made unpleasant if unplanned and poorly managed solutions are used to cope with these difficulties. A well-planned and efficient wastewater management system is essential for municipal management authorities to build a sustainable society. This may be accomplished via effective cooperation between residents and city officials, as well as strong governance. One

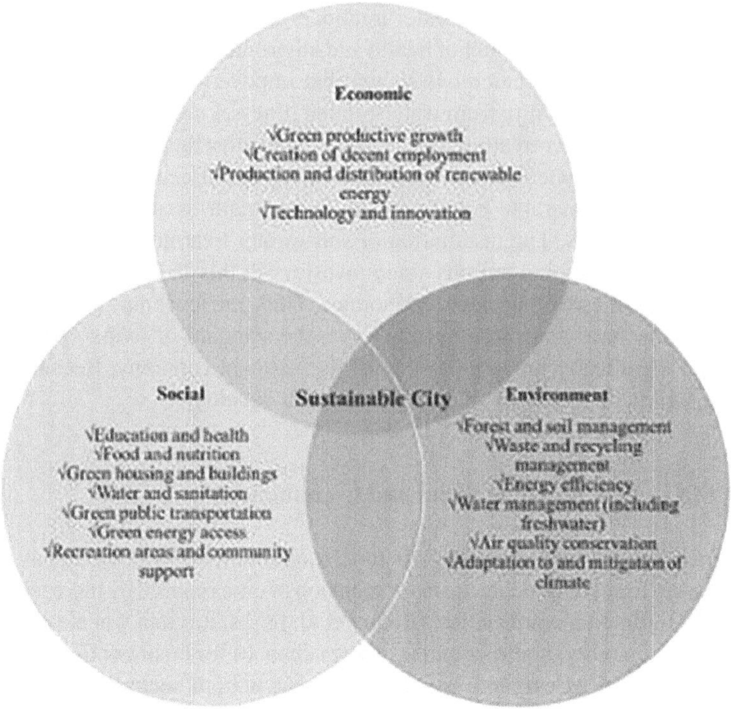

Fig. 5 Sustainable city pillars

of the main tenets of urban sustainability is expanding access to water (Fig. 5). The air quality will also improve as a result of this.

3.9.3 SDG 12: Responsible Consumption and Production

The majority of the natural resources that humans use up are non-reversible, meaning that once used, they cannot be transformed back into their original state. Over time, people will need to alter how they use these resources to make various products and consumer goods if we are to attain sustainable development and economic progress. Since water is a limited resource that people utilise for almost every aspect of their lives, from drinking to manufacturing, this is a very serious problem. Seventy percent of the world's freshwater comes from agricultural uses [86]. Humans significantly reduce the amount of freshwater available on Earth by polluting it with harmful waste and pollutants, which they then release into water streams. Many different industries and households also contribute to this problem by producing large amounts of highly polluted wastewater. Therefore, SDG 12 aims to promote the circular economy, which means that wastewater treatment will contribute to this goal by enhancing product recirculation and extending their useful life. To do this, we will recycle the water while removing valuable elements from the effluent [36, 87]. Using wastewater as an example, Kurniawan et al. [88] demonstrate how macrophytes may be used to create charcoal, animal feed, fertiliser, biofuel, and adsorbents.

According to Nhat et al. [89], half of the global population might face serious consequences due to freshwater shortages by the year 2050. In order to eradicate all of these water-related problems, it is critical to raise public awareness and persuade people to conserve this essential resource from pollution and waste. Businesses, industries, and other institutions should be encouraged to utilise water wisely to minimise water waste and to recycle polluted water for reuse after necessary treatment (De Francisci et al., (196). In order for regular people to have access to clean water and prevent its wasteful polluting, it is the duty of industrialised nations and NGOs to assist developing nations in constructing wastewater treatment facilities. Another advantage is that it protects the environment, especially water bodies, from contamination, which in turn eliminates the hazards of environmental pollution.

3.9.4 SDG 13: Climate Action

Increases in the rate of climate change are closely related to wastewater. One way this SDG plans to mitigate climate change is by promoting and regulating renewable energy sources [90]. Water is a major renewable energy source, therefore it's possible to chemically treat and recycle some of the wastewater to help provide green, cheap power. Developing nations, in particular, may lessen the impact of climate change by increasing their wastewater treatment capacity [91]. To reduce the demand for fresh and subterranean water resources, wastewater treated with chemicals to remove contaminants and other contaminants may be used to generate electricity. Dogan and Seker argue that progress in the renewable energy sector has the potential to significantly reduce pollution levels, ensuring a healthy and clean environment for all living things [92]. Seawater desalination is sometimes associated with additional environmental issues, which is why wastewater recycling is recommended [93]. Upcoming renewable energy projects may make better use of treated wastewater, which means less fresh water from rivers and lakes. Beyond this, there are more potential uses for the treated wastewater that was released into the environment, such as in farming, household cleaning, and washing machines.

Companies are compelled to increase their usage of renewable resources, as opposed to conventional power plants, as a result of the government's climate change policy. One way to meet climate change goals is to generate electricity from treated wastewater in an eco-friendly way [94]. If the treatment facility accounts for any and all greenhouse gas emissions that may occur there, this will lead to a decrease in total GHG emissions [95, 96]. For instance, major power plants and electrical manufacturing businesses may cut down on their CO2 emissions with the support of a growing renewable energy industry. Industries are increasingly concentrating on recycling wastewater as an alternative to discharging it into the earth and waterways. Acid rain and climate change are two problems that this programme aims to mitigate. In addition to this, it will aid in reducing the fast and unpredictable climate change while also protecting the ecology from pollutants. Less untreated wastewater, which includes sewage litter and other particles that might harm the environment, is one of the main aims of wastewater treatment strategies. Marine species are less likely to suffer from long-term environmental harm because to these rules. Successful treatment plans and procedures include discharging treated wastewater back into the environment. River flows are vital for conservation, fishing, and leisure activities, and this helps keep them going [97]. All firms, no matter how large or little, are encouraged by government climate change policies to engage in environmentally conscious activities that will benefit generations to come. Not only that, but SDG 13 is only one of several SDGs that benefit from the biogas business, which relies heavily on the wastewater treatment process [63].

3.9.5 SDG 14: Life Below Water

Everything residing in or around water, whether it be plants or animals, is considered part of the aquatic ecosystem. Water is home to many different kinds of reptiles and fish; thus, it is crucial that we do everything in our power to protect these creatures from harm. It is arguable that the two topics of wastewater management and marine life are closely intertwined. After being cleansed, water that has been polluted by humans is discharged into rivers, where it finally flows into the ocean. Without proper treatment, the water will be contaminated with chemicals and other contaminants that pose a significant threat to aquatic life [98]. Wastewater that has been recycled has the potential to kill marine life and other aquatic species if not properly handled. Wastewater treatment and filtration might render harmless the toxins and other contaminants that threaten marine life. It is crucial to treat wastewater for aquatic life since it may have long-term effects on aquatic plants and animals as well as any humans who come into contact with it [99].

In many regions of the globe, people eat fish and other marine creatures. The continuous release of wastewater into the ocean poses serious health risks to marine life. Humans may face even more serious health problems after consuming these animals that have been impacted by polluted water. That is to say, people may also develop serious health problems if they eat animals that have been exposed to this kind of effluent.

3.9.6 SDG 15: Life Land

Researchers have found a strong correlation between water scarcity and land degradation [100]. The capacity of healthy soil to store water is inherent. But when the land is damaged, this ability disappears. For example, the soil's quality has deteriorated due to inadequate irrigation management and massive amounts of drainage runoff, which hinders plant and tree development. Consequently, plans and regulations must be put in place to control the overflow of wastewater onto land and to protect the water and land resources. An environmental impact statement (EIS) has been prepared for the purpose of planning the proposal to upgrade the wastewater treatment process. An EIS is required by law for every wastewater treatment facility that generates more than 10,000 PE, according to the research. Following the rules for managing wastewater flow improves soil, amenity value, fisheries, tourism, and water quality. Hopefully, humans, animals, and other land-dwelling organisms will all benefit from wastewater treatment. The wastewater project at the global water awards went to the Paso Robles wastewater treatment facility in the USA. This is because it demonstrates the most innovative approach to minimising the plant's physical or ecological footprint, which has a significant influence on ecosystems and biodiversity [101].

3.9.7 SDG 16: Peace, Justice and Strong Institution

A significant and long-lasting reduction in corruption of water resources is indicative of effective anti-corruption strategies. Organisations can assess, prevent, decrease, and monitor bribery situations with the use of anti-bribery management systems [102]. Using the approved ISO 37001 Anti-bribery Management System, wastewater treatment companies can protect their standing with stakeholders. The certification is useful for hydropower generating and water treatment companies because it allows stakeholders to keep an eye out for and control the risk of any illegal activity related to the distribution of cheap hydro energy. The certification also proves that the vendors, subcontractors, agents, and distributors are committed to not bribing anybody. The power and utility sectors are treating and distributing wastewater in accordance with human rights standards [103]. In order to protect the interests of workers, customers, and the environment, several entities involved in hydropower development and wastewater treatment facilities have distinct characteristics. The idea of wastewater treatments has widespread community and stakeholder support, indicating that it is a reasonable and peaceful solution. Wastewater treatment project managers must adhere strictly to all agreements and regulations pertaining to sustainable development and growth if we are to live in a fair and peaceful world [104].

3.9.8 SDG 17: Partnerships for the Goals

Government tax policies that seek to minimise the consumption of both fresh water and wastewater will better safeguard aquatic ecosystems. There is an inverse relationship between the capacity of discharged wastewater and the government-developed taxes, which are in turn based on the water's ecological quality [105]. The government-approved levies that are meant to correct the market inefficiencies. Wastewater discharge and environmental pollution have been considerably reduced thanks to the authorised levies, which impact the actions of the polluting actors. Organisations are incentivized via taxes to improve their wastewater treatment facilities, which in turn leads to less land degradation and deforestation. Companies are obligated to pay ecotaxes, often called environmental taxes, as outlined in the Water

Act [106]. When it comes to paying taxes, there are no regulations regarding compensation or exchange. No matter how much money is made from wastewater treatment, businesses are still obligated by the tax policy to make improvements to the process so that the environment remains clean and healthy.

References

1. Capodaglio A, Callegari A, Cecconet D, Molognoni D (2017) Sustainability of decentralized wastewater treatment technologies. Water Pract Technol 12:463–477
2. Malik OA, Hsu A, Johnson LA, de Sherbinin A (2015) A global indicator of wastewater treatment to inform the sustainable development goals (SDGs). Environ Sci Policy 48:172–185
3. Hák T, Janoušková S, Moldan B (2016) Sustainable development goals: a need for relevant indicators. Ecol Indic 60:565–573
4. Stafford-Smith M, Griggs D, Gaffney O, Ullah F, Reyers B, Kanie N, Stigson B, Shrivastava P, Leach M, O'Connell D (2017) Integration: the key to implementing the sustainable development goals. Sustain Sci 12:911–919
5. Bebbington J, Unerman J (2018) Achieving the United Nations sustainable development goals account. Audit Account J 31:2–24. https://doi.org/10.1108/AAAJ-05-2017-2929
6. Costanza R, Daly L, Fioramonti L, Giovannini E, Kubiszewski I, Mortensen LF, Pickett KE, Ragnarsdottir KV, De Vogli R, Wilkinson R (2016) Modelling and measuring sustainable wellbeing in connection with the UN sustainable development goals. Ecol Econ 130:350–355
7. Prabhakaran D, Andersen TL, Markussen C, Madsen B, Lilholt H (2013) Tensile and compression properties of hybrid composites-a comparative study. In: Proceedings of the 19th international conference on composite materials (ICCM19), pp 1029–1035
8. Rashidi H, GhaffarianHoseini A, Sulaiman NMN, Tookey J, Hashim NA (2015) Application of wastewater treatment in sustainable design of green built environments: a review. Renew Sustain Energy Rev 49:845–856
9. Henriques J, Catarino J (2017) Sustainable value-an energy efficiency indicator in wastewater treatment plants. J Clean Prod 142:323–330
10. Kurian M (2017) The water-energy-food nexus: trade-offs, thresholds and transdisciplinary approaches to sustainable development. Environ Sci Policy 68:97–106
11. Zhang Q, Prouty C, Zimmerman JB, Mihelcic JR (2016) More than target 6.3: a systems approach to rethinking sustainable development goals in a resource-scarce world. Engineering 2:481–489
12. Singh KP, Mohan D, Sinha S, Dalwani R (2004) Impact assessment of treated/untreated wastewater toxicants discharged by sewage treatment plants on health, agricultural, and environmental quality in the wastewater disposal area. Chemosphere 55:227–255
13. Albolafio S, Marín A, Allende A, García F, Simón-Andreu PJ, Soler MA, Gil MI (2022) Strategies for mitigating chlorinated disinfection byproducts in wastewater treatment plants. Chemosphere 288(132583)
14. Guidelines for social life cycle assessment of products and organizations (2020). https://www.lifecycleinitiative.org/wp-content/uploads/2021/01/Guidelines-for-Social-Life-Cycle-Assessment-of-Products-and-Organizations-2020-22.1.21sml.pdf
15. Integrating SDGs into sustainability reporting (2022) https://www.globalreporting.org/public-policy-partnerships/sustainable-development/integrating-sdgs-into-sustainability-reporting/
16. Rafiaani P, Dikopoulou Z, Van Dael M, KuppensT, Azadi H, Lebailly P, Van Passel S (2020) Identifying social indicators for sustainability assessment of CCU technologies: a modified multi-criteria decision making. Soc Indic Res 147:15–44
17. Jarvis P (2020) Environmental technology for the sustainable development goals (SDGs), Taylor & Francis

18. Dilekli N, Cazcarro I (2019) Testing the SDG targets on water and sanitation using the world trade model with a waste, wastewater, and recycling framework. Ecol Econ 165(106376)
19. Zhou X (2018) SDG interlinkages analysis and applications for integrated policy making. JSTOR
20. Sims N, Kasprzyk-Hordern B (2020) Future perspectives of wastewater-based epidemiology: monitoring infectious disease spread and resistance to the community level. Environ Int 139
21. Shomar B, Dare A (2015) Ten key research issues for integrated and sustainable wastewater reuse in the middle east. Environ Sci Pollut Res 22:5699–5710
22. Hernández-Sancho F, Lamizana-Diallo B, Mateo-Sagasta J, Qadir M (2015) Economic valuation of wastewater: the cost of action and the cost of no action, United Nations Environment Programme (UNEP), Kenya
23. Renner M (2017) Wastewater and jobs the decent work approach to reducing untreated wastewater. Int Labour Org
24. Sayed ET, Abdelkareem MA, Alawadhi H, Olabi AG (2021) Enhancing the performance of direct urea fuel cells using Co dendrites. Appl Surf Sci 555(149698)
25. Abdelkareem MA, Sayed ET, Mohamed HO, Obaid M, Rezk H, Chae KJ (2020) Nonprecious anodic catalysts for low-molecular-hydrocarbon fuel cells theoretical consideration and current progress. Prog Energy Combust Sci 77(100805)
26. Alawadhi H, Abdelkareem MA, Hussain N, Wilberforce T, Sayed ET (2020) A composite of graphitic carbon nitride and Vulcan carbon as an effective catalyst support for Ni in direct urea fuel cells. J Taiwan Inst Chem Eng 116:160–168
27. Sayed ET, Abdelkareem MA, Bahaa A, Eisa T, Alawadhi H, Al-Asheh S, Chae KJ, Olabi AG (2021) Synthesis and performance evaluation of various metal chalcogenides as active anodes for direct urea fuel cells Renew. Sustain Energy Rev 150(111470)
28. Tawalbeh M, Al-Othman A, Singh K, Douba I, Kabakebji D, Alkasrawi M (2020) Microbial desalination cells for water purification and power generation: a critical review. Energy 209(118493)
29. Liu F, Moustafa H, Hassouna MSE D, He Z (2020) Resource recovery from wastewater can be an application niche of microbial desalination cells. Environ Int 142(105855)
30. Sayed ET, Abdelkareem MA (2017) Yeast as a biocatalyst in microbial fuel cell, old yeasts-new questions In Tech, pp 41–65
31. Mohamed HO, Obaid M, Sayed ET, Abdelkareem MA, Park M, Liu Y, Kim HY, Barakat NAM (2017) Graphite sheets as high-performance low-cost anodes for microbial fuel cells using real food wastewater Chem. Eng Technol 40:2243–2250
32. Kadier A, Jain P, Lai B, Kalil MS, Kondaveeti S, Alabbosh KFS, Abu-Reesh IM, Mohanakrishna G (2020) Biorefinery perspectives of microbial electrolysis cells (MECs) for hydrogen and valuable chemicals production through wastewater treatment. Biofuel Res J 7:1128–1142
33. Kadier A, Al-Shorgani NK, Jadhav DA, Sonawane JM, Mathuriya AS, Kalil MS, Hasan HA, Alabbosh KFS (2020) Microbial electrolysis cell (MEC) an innovative waste to bioenergy and value-added by-product technology. Bioelectrosynthesis: principles and technologies for value-added products, Wiley? VCH Verlag GmbH & Co. KGaA, pp 95–128
34. Hukari S, Hermann L, Nättorp A (2016) From wastewater to fertilisers–technical overview and critical review of European legislation governing phosphorus recycling. Sci Total Environ 542:1127–1135
35. Løes AK (2021) Feeding the reactors: potentials in re-cycled organic fertilisers. Org Agric 11:245–250
36. Hussain F, Shah SZ, Ahmad H, Abubshait SA, Abubshait HA, Laref A, Manikandan A, Kusuma HS, Iqbal M (2021) Microalgae an ecofriendly and sustainable wastewater treatment option: biomass application in biofuel and bio-fertilizer production. Review Renew Sustain Energy Rev 137(110603)
37. Minale M, Worku T (2014) Anaerobic co-digestion of sanitary wastewater and kitchen solid waste for biogas and fertilizer production under ambient temperature: waste generated from condominium house. Int J Environ Sci Technol 11:509–516

38. Priyanath HMS, Premaratne S, Yoosuf A, Maurice D (2018) Technical efficiency for tea smallholder farmers under UTZ certification system in Sri Lanka: a stochastic frontier approach SEISENSE. J Manag 1:1–19
39. Jaramillo MF, Restrepo I (2017) Wastewater reuse in agriculture: a review about its limitations and benefits Sustainability 9:1734
40. Delanka-Pedige H, Munasinghe-Arachchige S, Abeysiriwardana-Arachchige I, Nirmalakhandan N (2020) Wastewater infrastructure for sustainable cities: assessment based on UN sustainable development goals (SDGs). Int J Sustain Dev World Ecol 28:1–7. https://doi.org/10.1080/13504509.2020.1795006
41. Valipour M, Singh VP (2016) Global experiences on wastewater irrigation: challenges and prospects balanced Urban development: options and strategies for liveable cities.Springer, Cham, pp 289–327
42. Behera BK, Rout PK, Behera S (2019) Water, energy and food security: pillars for zero hunger in move towards zero hunger. Springer, pp 37–60
43. Zohar I, Forano C (2021) Phosphorus recycling potential by synthetic and waste materials enriched with dairy wastewater: a comparative physicochemical study. J Environ Chem Eng 9(106107)
44. Moustafa MT (2017) Removal of pathogenic bacteria from wastewater using silver nanoparticles synthesized by two fungal species. Water Sci 31:164–176
45. Zhu H, Han Y, Xu C, Han H, Ma W (2018) Overview of the state of the art of processes and technical bottlenecks for coal gasification wastewater treatment. Sci Total Environ 637:1108–1126
46. Tahir MB, Nawaz T, Nabi G, Sagir M, Khan MI, Malik N (2020) Role of nanophotocatalysts for the treatment of hazardous organic and inorganic pollutants in wastewater. Int J Environ Anal Chem 102:1–25. https://doi.org/10.1080/03067319.2020.1723570
47. McCall C, Wu H, Miyani B, Xagoraraki I (2020) Identification of multiple potential viral diseases in a large urban center using wastewater surveillance. Water Res 184(116160)
48. Salgot M, Folch M (2018) Wastewater treatment and water reuse. Curr Opin Environ Sci Health 2:64–74
49. Fan F, Xu R, Wang D, Meng F (2020) Application of activated sludge for odor control in wastewater treatment plants: approaches, advances and outlooks. Water Res 181(115915)
50. Jain K, Patel AS, Pardhi VP, Flora SJS (2021) Nanotechnology in wastewater management: a new paradigm towards wastewater treatment. Molecules 26:1797
51. Shakir E, Zahraw Z, Al-Obaidy AHM (2017) Environmental and health risks associated with reuse of wastewater for irrigation Egypt. J Pet 26:95–102
52. Cooper JA, Loomis GW, Kalen DV, Amador JA (2015) Evaluation of water quality functions of conventional and advanced soil-based onsite wastewater treatment systems. J Environ Qual 44:953–962
53. Herrera JA (2020) Melian sustainable wastewater treatment systems (2018–2019) Multidisciplinary Digital Publishing Institute
54. Angelakis AN, Zheng XY (2015) Evolution of water supply, sanitation, wastewater, and stormwater technologies globally multidisciplinary digital publishing institute
55. Dickin S, Bayoumi M, Giné R, Andersson K, Jiménez A (2020) Sustainable sanitation and gaps in global climate policy and financing. NPJ Clean Water 3:1–7
56. Lu Y, Song S, Wang R, Liu Z, Meng J, Sweetman AJ, Jenkins A, Ferrier RC, Li H, Luo W (2015) Impacts of soil and water pollution on food safety and health risks in China. Environ Int 77:5–15
57. Winkler MS, Jackson D, Sutherland D, Lim JMU, Srikantaiah V, Samuel Fuhrimann KM (2017) Sanitation safety planning as a tool for achieving safely managed sanitation systems and safe use of wastewater. WHO South East Asia J Public Health 6:34–40
58. UIEA, Global energy review (2020), Ukraine. https://www.iea.org/countries/ukraine. Accessed 2020–09–10
59. Heidrich ES, Curtis TP, Dolfing J (2011) Determination of the internal chemical energy of wastewater. Environ Sci Technol 45:827–832

60. Yilmaz I, Gumus M (2017) Investigation of the effect of biogas on combustion and emissions of TBC diesel engine. Fuel 188: 69–78
61. Abdelkareem MA, Tanveer WH, Sayed ET, Assad MEH, Allagui A, Cha SW (2019) On the technical challenges affecting the performance of direct internal reforming biogas solid oxide fuel cells. Renew Sustain Energy Rev 101:361–375
62. Witte J, Calbry-Muzyka A, Wieseler T, Hottinger P, Biollaz SM, Schildhauer TJ (2019) Demonstrating direct methanation of real biogas in a fluidised bed reactor. Appl Energy 240:359–371
63. Obaideen K, Abdelkareem MA, Wilberforce T, Elsaid K, Sayed ET, Maghrabie HM, Olabi A (2022) Biogas role in achievement of the sustainable development goals: evaluation, challenges, and guidelines. J Taiwan Inst Chem Eng 131(104207)
64. World Bank, wastewater a resource that can pay dividends for people, the environment, and economies, says World Bank. Available at. https://www.worldbank.org/en/news/press-release/2020/03/19/wastewater-a-resource-that-can-pay-dividends-for-people-the-environment-and-economies-says-world-bank. Accessed 17 March 2021
65. Timmis K, De Lorenzo V, Verstraete W, Ramos JL, Danchin A, Brüssow H, Singh BK, Timmis JK (2017) The contribution of microbial biotechnology to economic growth and employment creation. Microb Biotechnol 10:1137–1144
66. Shi Z, She Z, Chiu YH, Qin S, Zhang L (2021) Assessment and improvement analysis of economic production, water pollution, and sewage treatment efficiency in China. Socioecon Plann Sci 74(100956)
67. Abu-Ghunmi D, Abu-Ghunmi L, Kayal B, Bino A (2016) Circular economy and the opportunity cost of not 'closing the loop' of water industry: the case of Jordan. J Clean Prod 131:228–236
68. Avula RY, Nelson HM, Singh RK (2009) Recycling of poultry process wastewater by ultrafiltration. Innov Food Sci Emerg Technol 10:1–8
69. Liu S, Li H, Daigger GT, Huang J, Song G (2022) Material biosynthesis, mechanism regulation and resource recycling of biomass and high-value substances from wastewater treatment by photosynthetic bacteria: a review. Sci Total Environ 820(153200)
70. Li C, Yu J, Li W, He Y, Qiu Y, Li P, Wang C, Huang F, Wang D, Gao S (2018) Immobilization, enrichment and recycling of Cr(VI) from wastewater using a red mud/carbon material to produce the valuable chromite ($FeCr_2O_4$). Chem Eng J 350:1103–1113
71. Mohamed HO, Talas SA, Sayed ET, Park SG, Eisa T, Abdelkareem MA, Fadali OA, Chae KJ, Castaño P (2021) Enhancing power generation in microbial fuel cell using tungsten carbide on reduced graphene oxide as an efficient anode catalyst material. Energy 229(120702)
72. Sayed ET, Abdelkareem MA, Alawadhi H, Elsaid K, Wilberforce T, Olabi AG (2021) Graphitic carbon nitride/carbon brush composite as a novel anode for yeast-based microbial fuel cells. Energy 221(119849)
73. Sayed ET, Alawadhi H, Olabi AG, Jamal A, Almahdi MS, Khalid J, Abdelkareem MA (2021) Electrophoretic deposition of graphene oxide on carbon brush as bioanode for microbial fuel cell operated with real wastewater. Int J Hydrog Energy 46:5975–5983
74. Liang D, Zhang L, He W, Li C, Liu J, Liu S, Lee HS, Feng Y (2020) Efficient hydrogen recovery with CoP-NF as cathode in microbial electrolysis cells. Appl Energy 264(114700)
75. Wilberforce T, Sayed ET, Abdelkareem MA, Elsaid K, Olabi AG (2021) Value added products from wastewater using bioelectrochemical systems: current trends and perspectives. J Water Process Eng 39(101737)
76. Olabi AG, Wilberforce T, Sayed ET, Elsaid K, Rezk H, Abdelkareem MA (2020) Recent progress of graphene based nanomaterials in bioelectrochemical systems. Sci Total Environ 749(141225)
77. Sayed ET, Shehata N, Abdelkareem MA, Atieh MA (2020) Recent progress in environmentally friendly bio-electrochemical devices for simultaneous water desalination and wastewater treatment. Sci Total Environ 748(141046)
78. Hua T, Li S, Li F, Zhou Q, Ondon BS (2019) Microbial electrolysis cell as an emerging versatile technology: a review on its potential application, advance and challenge. J Chem Technol Biotechnol 94:1697–1711

79. Qu J, Wang H, Wang K, Yu G, Ke B, Yu HQ, Ren H, Zheng X, Li J, Li WW (2019) Municipal wastewater treatment in China: development history and future perspectives. Front Environ Sci Eng 13:1–7
80. Ren B, Zhao Y, Lyczko N, Nzihou A (2019) Current status and outlook of odor removal technologies in wastewater treatment plant. Waste Biomass Valoriz 10:1443–1458
81. Reymond P, Wahaab RA, Moussa MS, Lüthi C (2018) Scaling up small scale wastewater treatment systems in low-and middle-income countries: an analysis of challenges and ways forward through the case of Egypt. Utilit Policy 52:13–21
82. Valta K, Aggeli E, Papadaskalopoulou C, Panaretou V, Sotiropoulos A, Malamis D, Moustakas K, Haralambous KJ (2015) Adding value to olive oil production through waste and wastewater treatment and valorisation: the case of Greece Waste Biomass Valoriz 6:913–925
83. Street R, Malema S, Mahlangeni N, Mathee A (2020) Wastewater surveillance for Covid-19: an African perspective. Sci Total Environ 743(140719)
84. Yuan Q, McIntyre N, Wu Y, Liu Y, Liu Y (2017) Towards greater socio-economic equality in allocation of wastewater discharge permits in China based on the weighted Gini coefficient. Resour Conserv Recycl 127:196–205
85. El Zein Z, Abdou A, Abd ElGawad I (2016) Constructed wetlands as a sustainable wastewater treatment method in communities. Procedia Environ Sci 34:605–617
86. Abobatta W (2018) Impact of hydrogel polymer in agricultural sector. Adv Agric Environ Sci Open Access 1:59–64
87. Guerra-Rodríguez S, Oulego P, Rodríguez E, Singh DN, Rodríguez-Chueca J (2020) Towards the implementation of circular economy in the wastewater sector: challenges and opportunities. Water 12:1431
88. Kurniawan SB, Ahmad A, Said NSM, Imron MF, Abdullah SRS, Othman AR, Purwanti IF, Hasan HA (2021) Macrophytes as wastewater treatment agents: nutrient uptake and potential of produced biomass utilization toward circular economy initiatives. Sci Total Environ 790(148219)
89. Nhat PVH, Ngo H, Guo W, Chang S, Nguyen D, Nguyen P, Bui X, Zhang X, Guo J (2018) Can algae-based technologies be an affordable green process for biofuel production and wastewater remediation?. Bioresour Technol 256:491–501
90. Griggs D, Nilsson M, Stevance A, McCollum D (2017) A guide to SDG interactions: from science to implementation international council for science, Paris
91. Mara D (2013) Domestic wastewater treatment in developing countries Routledge
92. Dogan E, Seker F (2016) The influence of real output, renewable and non-renewable energy, trade and financial development on carbon emissions in the top renewable energy countries. Renew Sustain Energy Rev 60:1074–1085
93. Parkinson S, Krey V, Huppmann D, Kahil T, McCollum D, Fricko O, Byers E, Gidden MJ, Mayor B, Khan Z (2019) Balancing clean water-climate change mitigation trade-offs. Environ Res Lett 14(014009)
94. Quaschning VV (2019) Renewable energy and climate change John Wiley & Sons
95. Mamais D, Noutsopoulos C, Dimopoulou A, Stasinakis A, Lekkas T (2015) Wastewater treatment process impact on energy savings and greenhouse gas emissions. Water Sci Technol 71:303–308
96. Mannina G, Ekama G, Caniani D, Cosenza A, Esposito G, Gori R, Garrido-Baserba M, Rosso D, Olsson G (2016) Greenhouse gases from wastewater treatment-a review of modelling tools. Sci Total Environ 551:254–270
97. Gardner M, Jones V, Comber S, Scrimshaw MD, Coello-Garcia T, Cartmell E, Lester J, Ellor B (2013) Performance of UK wastewater treatment works with respect to trace contaminants. Sci Total Environ 456:359–369
98. Spiller M, Vreeburg JH, Leusbrock I, Zeeman G (2015) Flexible design in water and wastewater engineering-definitions, literature and decision guide. J Environ Manag 149:271–281
99. Thines R, Mubarak N, Nizamuddin S, Sahu J, Abdullah E, Ganesan P (2017) Application potential of carbon nanomaterials in water and wastewater treatment: a review. J Taiwan Inst Chem Eng 72:116–133

100. Karamesouti M, Detsis V, Kounalaki A, Vasiliou P, Salvati L, Kosmas C (2015) Land-use and land degradation processes affecting soil resources: evidence from a traditional Mediterranean cropland (Greece) Catena 132:45–55
101. Waqas S, Bilad MR, Man ZB, Klaysom C, Jaafar J, Khan AL (2020) An integrated rotating biological contactor and membrane separation process for domestic wastewater treatment Alex. Eng J 59:4257–4265
102. Mareai BM, Fayed M, Aly SA, Elbarki WI (2020) Performance comparison of phenol removal in pharmaceutical wastewater by activated sludge and extended aeration augmented with activated carbon. Alex Eng J 59:5187–5196
103. Khalaf AH, Ibrahim WA, Fayed M, Eloffy MG (2021) Comparison between the performance of activated sludge and sequence batch reactor systems for dairy wastewater treatment under different operating conditions, Alex. Eng J 60:1433–1445
104. Cohen A (2011) Values and psychological contracts in their relationship to commitment in the workplace. Career Dev Int 16:646–667. https://doi.org/10.1108/13620431111187272
105. Stahel WR (2013) Policy for material efficiency-sustainable taxation as a departure from the throwaway society. Philos Trans R Soc A 371(20110567)
106. Gallego Valero L, Moral Pajares E, Román Sánchez IM (2018) The tax burden on wastewater and the protection of water ecosystems in EU countries. Sustainability 10:212

Characteristics of Wastewater

Aftabuz Zaman

Abstract All known forms of life depend critically on water, a naturally occurring resource. Clean water is in high demand right now, but it's also under danger from or dependent on a number of pollution sources, some of which are listed below: Garbage from homes and farms, as well as all types of soaps and detergents. Radioactive materials, thermal pollution, toxic metals, fertilisers, industrial waste, etc.

All known forms of life depend critically on water, a naturally occurring resource. Clean water is in high demand right now, but it's also under danger from or dependent on a number of pollution sources, some of which are listed below: Garbage from homes and farms, as well as all types of soaps and detergents. Radioactive materials, thermal pollution, toxic metals, fertilisers, industrial waste, etc.

All of the sewage that normally makes up the Wastewater characteristics locality—from houses and other residential buildings to businesses and other institutions—is considered domestic flow. It is necessary to calculate more expressive values from significant point sources independently and then add them to the global value. Typically, water consumption in a given area is taken into account while planning the flow of residential sewage. The conventional method for estimating water consumption is to take the design population into account and then add a value to the average daily per capita water usage.

1 Sewage and Domestic Wastewater

Pollution of water sources is mostly caused by municipal water. There are trace amounts of harmful metals including Cu, Cr, Zn, Mn, Pb, and Ni in household sewage. Because of the organic materials it contains, sewage puts a demand on the

A. Zaman (✉)
School of Agricultultural Sciences, Sister Nivedita University, Kolkata, India
e-mail: profazaman@gmail.com

B. G. Gupta (ed.), *Urban Development with Treated Recycled and Reuse of Wastewater: Sustainable Management*, Urban Sustainability,
https://doi.org/10.1007/978-981-96-0857-7_5

receiving waters for oxygen. Fatty acids, esters, amino acids, amides, amino sugars, and proteinaceous amines are all usually part of organic matter. Released partially digested wastes containing some biochemical oxygen demand (BOD) are the primary source of organic pollution in water.

Sewage in most Municipalities receives no treatment before they are discharged into water bodies.

2 Agriculture Discharges

Water supplies are reportedly heavily polluted by plant nutrients, pesticides, herbicides, fertilisers, agricultural wastes, manure slurry, sediments, silage drainage, plant and animal debris, and soil erosion, which primarily consists of inorganic components. Soil amendments using NPK fertilisers, which comprise nitrate and phosphate, are a staple of contemporary farming techniques. A portion of them end up in waterways due to runoff from precipitation, irrigation, and drainage, where they wreak havoc on aquatic ecosystems. A disturbance in the water's nitrogen and phosphorus balance, which in turn affected plant development, was caused by the overuse of plant fertilisers. Receiving water's BOD is increased by organic waste.

Organic waste from cattle, antibiotics, silage effluents, processing waste from plantation crops, and pesticides and fertilisers utilised in agriculture can also contaminate surface water and groundwater.

3 Industrial Wastewater

In the case of the Textile industry different processes comprise degumming, de-sizing, rummage, lightening, staining, printing, and making. The process of degumming involves removing the gum from silk fibres. Gulp, face, ability, and sensation are all improved after gum removal. To prepare the fibre for staining, it is left on top of the fibre with wax, which acts as a protective coating.

Desizing entails injecting the fibre with a desizing chemical, which then dissolves or reduces the amount of material and removes any damaged goods.

The first step in bleaching textiles—with or without chemicals—is scouring, which can be done at room temperature or at appropriate higher temperatures with other appropriate wetting agents, alkali, and so forth. By removing the waxes and pectins, scouring makes the fabric seem hydrophilic, meaning it absorbs water well.

Bleaching is a process that depolarizes grease material, making it easier to treat subsequently. Threads, yarn, and textiles are refined goods that may be coloured through the process of dyeing.

Dyeing is carried out in a specific chemical solution that contains pigments. Printing is the process of applying colour to fibre according to predetermined patterns or specifications. The colour is fixed with the thread in printed fabrics to prevent them from becoming damp or resistant.

In textile production, "finishing" refers to the steps taken after "staining" the thread or fabric to improve the aesthetics, functionality, or durability of the final product.

Unplanned growth of textile units impacts the degradation of the ecosystem. Surface water, groundwater, soil, agricultural crops and vegetables, air quality, and most importantly, human health are all negatively impacted by the discharge of effluents from textile bleaching and dyeing facilities [1]. Many of such Bleaching & Dyeing (B&D) units lack effluent treatment plant and the untreated wastewater is released to the canals, open sewers and indeed the nearby land. The wastewater discharged from textile units consists of a cocktail of toxic chemicals and dyes applied in the bleaching and dyeing operations.

3.1 Physicochemical Parameters of Wastewater

3.1.1 Wastewater pH

The pH is an important indicator for determining the nature of wastewater, and surface water and thus plays a significant aspect in environmental and chemical engineering studies. In textile productions, the pH of wastewater developed from bleaching and dyeing processes filled with diverse chemicals and dyes varies from 5.5 to 9. Studies have reported pH values varying from 6.5 to 8.9 at Karur district, Tamilnadu, India in 2005 for a similar situation of groundwater pollution [1].

3.1.2 Water Alkalinity

The reaction of bodies of water to acid loads can be better understood using this method. As a result of human illnesses, the alkalinity of rivers and streams can alter over time [2]. A substance's alkalinity is defined as its ability to reduce the acidity of a solution to a level where carbonate or bicarbonate is no longer produced [4] Bleaching and dyeing machines produce effluent with an alkalinity level ranging from 325 to 580 mg/L. Industrial surface water with an alkalinity level between 310 and 460 mg/L, far higher than the 250 mg/L limit for effluent [5, 6].

3.1.3 Total Dissolved Solids (TDS)

Trace amounts of inorganic and biological components floating in a colloidal solution are quantified as total dissolved solids (TDS). It is an indication of the behaviour of a vast variety of chemical contaminants and is a feature of drinking water.

Sodium, potassium, calcium, phosphates, nitrates, and chlorides are the most common chemical components, which are frequently supplemented by surface water, rainwater, and agricultural runoffs. In place of a thousand or fewer particles, big enough to form a soluble micro granule, the chemicals can be cations, anions, molecules, or agglomerations. A secondary water quality criterion of 500 mg/L has been set by the United States to ensure that drinking water is tasty [3]. Total suspended solids (TSS) are distinct from total dissolved solids (TDS) because TSS remain in solution even after passing through a two-micrometer screen. Materials of varying sizes that cannot be suspended or dissolved in a holding tank exempt from the act are referred to as "settable solids"; this phrase does not include TDS or TSS [4]. Larger particles or molecules that are insoluble can be considered settable solids.

Since 500 mg/L is the aesthetic threshold for total dissolved solids (TDS) in drinking water (as established by [5]), nobody cares about the colour, taste, or smell of the water, and the quantity is far lower than what's required to cause harm. Generally speaking, TDS levels of 1000 mg/L are not harmful to marine environments that contain a variety of fish species [6]. Discharges from various types of industrial activity have different total dissolved solids. The concentration of total dissolved solids (TDS) in wastewater discharged by the textile industry ranges from 3050 to 5060 mg/L. According to IS: 10,500, the maximum permissible concentration of TDS is 500 mg/L (BIS, 2012). The concentrations of total dissolved solids (TDS) in surface water and groundwater, respectively, in regions affected by B&D activities vary between 1660 and 2778 mg/L and 434 and 4700 mg/L. [7].

3.1.4 Biological Oxygen Demand (BOD)

The total amount of oxygen needed by marine microbes is known as biochemical oxygen demand (BOD). The enzymes that the native microbes have access to, the temperature, and the amount of nutrients all have a role. Researchers have found that after 20 days, 90% of the total BOD is calculated, and 68% in 5 days, all using the dissolved oxygen that is accessible [8]. The BOD level in polluted rivers can range from 2 to 8 mg/L. Bod readings more than 8 mg/L indicate potential river pollution. A reading of 20 mg/L or below would be indicative of municipal sewage treated using a three-stage procedure. While typical BOD values in raw sewage in the US and Europe are approximately 600 mg/L, they do vary among regions. According to research, the biological oxygen demand (BOD) in wastewater from B&D units may be as high as 244 mg/L (13) and as low as 12–40 mg/L (14, 15) and 1.5–4.00 mg/L (16, 17), respectively, in the locations where these units release their effluent.

3.1.5 Chemical Oxygen Demand (COD)

An indicator of how much oxygen water can take in through processes like organic molecule disintegration and inorganic chemical oxidation is the chemical oxygen demand. Samples of wastewater or naturally polluted waterways are used to evaluate COD. These samples are contaminated by wastes from homes or businesses". The maximum allowable concentration of COD for industrial effluent is 250 mg/L, according to the CPCB. Contaminated surface water in areas where textile B&D is prevalent has a COD level ranging from 46 to 943 mg/L, whereas COD levels in textile industry effluent range from 250 to 979 mg/L [7] (18).

3.1.6 Dissolved Oxygen (DO)

An essential and well-structured indicator of water quality is dissolved oxygen. Oxygen influences a wide range of other biochemical and aesthetic water quality parameters, including smell, clarity, and flavour. It is absolutely necessary for all kinds of aquatic life, including fish, plants, and bacteria. Every living thing has its own specific requirement for the amount of dissolved oxygen. Oxygen levels should be between 4 and 15 mg/L for shallow-water fish, but between 1 and 6 mg/L for bottom feeders, prawns, worms and crabs (19).

In order to preserve a variety of aerobic life forms, natural stream cleaning processes need adequate oxygen levels. Water life is endangered when the dissolved oxygen content falls below 5.0 mg/L. The level of stress increases as the quantity of dissolved oxygen decreases. Heavy fish mortality may occur if oxygen levels stay below 1–2 mg/L for more than a few hours.

Canal water that receives effluent from the textile industry contains dissolved oxygen levels ranging from 0.90 mg/L to 2.6 mg/L (20), while the effluent itself has no dissolved oxygen.

4 Origin of Chemical Contaminants Contributed from B&D Units

There are many different types of textile chemicals used in the process, beginning with the pre-treatment and continuing through the finishing stages. Scour works, tiny foam scour works, wetting agents, sequestering agents, mercerizing agents, peroxide stabilisers, peroxide killers, and neutralizers are among them. In this mixture, you will find sodium hypochloride, sulfuric acid, a bleaching agent, fifty percent hydrogen peroxide, thickeners, folds, stain removers, anti-back staining agents, reactive softeners, cationic softeners flakes/paste, non-ionic softeners flakes/paste, anionic softeners flakes/paste, and cold water soluble softener flakes.

There is a long series of chemical and non-chemical procedures that textiles must undergo throughout manufacture. Fabric quality, printing, dying, and pre-treatment are all steps in the textile finishing process. Biocides, fire retardants, water repellents, and warp lengths are three examples of the specialised chemicals used in textile chemical processes. A few examples of these are starch, sulfonated oils, waxes, emulsified oils and lubricants, and basic chemical combinations like surfactants. Yarn making, fabric pre-treatment and finishing, textile laminating and painting, and a plethora of other services all necessitate more than sixty different types of textile chemical products.

In B& D effluent the anion forms the growth in the tendency as Cl- > HCO_{3-} > NO_{3-} > SO_4 2- > F - > PO_4—and cations form the pattern of Na > Ca > Mg > K (21). When industrial wastes like dyes and bleaches are disposed of in either liquid or solid form in bodies of water or on land, they eventually seep into the groundwater system, where they endanger ecosystems and human health (22). Formaldehyde, heavy metals, allergic dyes, finishing compounds (such as softeners, fire retardants, and water repellant), pesticides, shippers, pentachlorophenol, biocides, and colours cleaving carcinogenic amines are among the most harmful contaminants found in textiles. Toxins that humans are particularly vulnerable to include heavy metals such as cadmium, chromium, nickel, arsenic, lead, and beryllium [9].

Nitrate (NO_3) and its uses in textiles cleaning

Ammonium nitrate and zinc nitrate catalysts are needed for lightening and dyeing activities. The anions (NO_{3-}) formed owing to the use of different chemicals in the wet processing (cleaning) of textile cloth. The quantity of NO_3 in surface water contaminated with B&D wastewater at Perur (Tamilnadu) is described as 32.42 mg/L [10]. sets the admissible limit of nitrate (NO_3) in wastewater as 45 mg/L. But it observed that nitrate (NO_3) in untreated wastewater varies from 35 mg/L to 65 mg/L in several textile and other industrial studies. It also observed that the NO_3 in polluted surface water varies from 6 mg/l to 15 mg/L in some textile and industrial operations.

Heavy metals

The effluent from textile B&D units consists of a combination of different heavy metals. They get moved on to canals carrying wastewater, to cut off surface waters, and groundwater and indeed intensified in soils and plants making such wastewater. There are a wide range of impacts on plant development caused by heavy metal absorption and growth. Because Pb, Ni, Cd, and Cr are toxic, they may affect the soil's productivity and water quality as well as fruits, vegetables, and crops [11]. Humans, animals, and aquatic life in water bodies are greatly concerned about the absorption of heavy metals by plants from polluted soil and water. Vegetables, fruits, and plants become contaminated with metals as a result of environmental medium transfer, which can be harmful to human health.

Lead its uses in textiles and toxic effects

Dyeing processes often use lead. Textile dyeing makes use of a wide variety of basic dyes, including reactive, direct, vat, dispersion, azo, and mordant dyes. The colours all include heavy metals. Lead is known by several names in the textile industry for its many uses, including textile dyeing (lead acetate), lead salt preparation (lead chloride), lead moly date (dyestuff pigments), lead nitrate (oxidizer and mordant in dyeing), etc. Textile wastewater containing Pb in mg/L varies from 0.05 to 3 mg/L against the permissible limit as determined by IS: 10,500(2012) as

0.05 mg/L. The study by [12] at Tirupur (Tamilnadu) found Pb concentration as 1.30 mg/L in wastewater. Heavy metals like Pb are toxic non-biodegradable chemicals employed in chemical treating that are troublesome to separate from wastewater and may require tertiary and further treatment processes, and a decline to do so appears in natural threats. Heavy metals, like Pb present in textile industry wastewater, are not biodegradable, hence, they pass on from wastewater to surface water, surface water to groundwater and soil, soil to fruits and vegetables, and to the human body. Pb accumulates in vital organs in the body and over time provokes it, leading to various illnesses. Thus, untreated textile effluent can be harmful to both aquatic and terrestrial existence by changing the natural ecosystem and instigating long-term health impacts. A similar type of study on textile wastewater containing dyes and heavy metal (Pb) has shown metal accumulation and its impact on the environment.

Chromium, its uses in textiles, and its toxic effects

Chromium is used in textile dyes, paints, and leather tanning compounds. The textile industry makes use of a wide variety of dyes, including basic dyes, fibre reactive dyes, acid dyes, acid levelling dyes, acid milling dyes, acid dye colours, and mordant dyes. These dyes are all based on Cr. Toxic metal (Cr) is released through textile wastewater to the environment. As the metal Cr is non-biodegradable, it transfers from different water forms to soil, vegetable, fruits, and human health. Textile industries discharge Cr having concentration levels of 3–5 mg/L in wastewater against the permissible limit of 0.05 mg/L as per IS: 10,500 (2012). Cr concentration in the soil becomes 50 mg/kg owing to irrigation by textile wastewater or contaminated surface water. The study at Sanganer textile hub (Rajasthan) showed Cr concentration as 3—5 mg/L). The toxic effect due to Cr is high for the human being. It affects different organs of the human body and creates symptoms of illness. Similar studies showed a higher Cr concentration in textile wastewater.

Aluminum–its uses and toxic effects

Due to its great sensitivity, naturally occurring types of aluminium metal are exceedingly rare and found only in very reduced conditions. More than 270 distinct metals really contained it.

When consumed in large quantities, aluminium has been linked to changes in how the blood–brain barrier functions in disturbed humans. Contact dermatitis, gastrointestinal disorders, vomiting, and other symptoms can occur in a tiny number of persons who are hypersensitive to aluminium. Aluminium enhances the expression of genes associated with oestrogen in human breast cancer cells that have been grown in a controlled environment. Because of their estrogenic properties, these salts are now known as metal oestrogens.

Few studies have shown that the aluminium in deodorants actually irritates the skin. Some people are worried about its presence in dietary supplements, antiperspirants, and colours (like aluminium lake). But there isn't much proof that healthy individuals are at danger from the typical aluminium risk. It is possible that food-borne aluminium (Al) absorbs more than water-borne aluminium (Al).

Cadmium—its uses and toxic effects

The textile dyeing process makes use of a variety of natural dyes, including cadmium (Cd), which is present in azo dyes, direct dyes, and mordant dyes. The colours all include heavy metals. Kidney disease, early atherosclerosis, hypertension, and cardiovascular disorders are among the many conditions linked to cadmium exposure. Some experimental investigations have demonstrated that cadmium, an endocrine disruptor, can interact with many hormonal signalling pathways. To influence signal transduction in the oestrogen and MAPK signalling pathways, cadmium can bind to the oestrogen receptor alpha at low concentrations [10]. Cis concentration was found to be 0.098 mg/L [10, 13], which is above than the allowable limit according to IS:10,500 (2012), in a research conducted on textile wastewater in Hardwar, India. In accordance with BIS (2012), 0.005 mg/L is the maximum allowable concentration of Cd. There was an increased concentration of Cd in textile effluent, according to similar investigations [10].

4.1 Bleaching and Dyeing Process in Textiles

Bleach is a chemical that whitens cloth. Chlorine is the base for the most common bleaches. Coloured substances used in textiles that mimic the appearance of their substrate are called pigments. To increase the dye's fastness on the cloth, a caustic may be necessary when applying the dye in a colourless solution.

Synthetic dye in textile processing

The ingredients for synthetic colours are metals found in nature and by-products of the petroleum industry. In the time after, hundreds upon thousands of synthetic colours were created. Dye classes based on synthetic characteristics and solubility.

Fibres like wool, nylon, and modified acrylic are dyed using acid stains, which are anionic dyes that dissolve in water, in neutral to acid dye baths.

Acrylic fibre, and on rare occasions wool and silk, are dyed with basic dyes, which are cationic dyes that dissolve in water.

At or around the boiling point, either sodium chloride (NaCl), sodium sulphate (Na_2SO_4), or sodium carbonate (Na_2CO_3) is added to a neutral or alkaline dye tub to conduct direct or substantive dyeing. Fabrics such as cotton, paper, leather, wool, silk, and nylon can be dyed directly. In addition to being biological stains, they served as markers of pH.

A caustic is necessary for mordant dyes in order to restore the dye's resistance to water, light, and sweat. The final colour was considered when selecting the mordents [13].

Dyeing fibres using vat dyes is not possible since they are insoluble in water. Indigo, the first vat dye, is responsible for denim's colour.

A chromophore coupled to a reactive substituent that can react with the fibre substrate is necessary for reactive dyes. For home or studio dyeing of cotton and other cellulose fibres, reactive dyes are the way to go.

These water-insoluble disperse dyes were developed for use in cellulose acetate dyeing processes; the pigments are ground with a dispersion agent and either marketed as a paste or as dust after spray drying. Although polyester is their primary use, they are also capable of dyeing nylon, acrylic, and cellulose triacetate.

An insoluble azo dye is created onto or inside the fibre during the azoic dyeing process. A fibre can be treated with diazo and coupling factors simultaneously to achieve this effect. The poisonous chemicals employed in this procedure of dying cotton are reducing its worth.

One cheap way to dye cotton a deep shade of colour is with sulphur dyes. Cloth is dyed by immersing it in a mixture of a biological substance, a nitrophenol derivative, and sulphide or polysulfide while it is heated [9].

5 Microbial Contamination

The textile effluent itself is not polluted with the microbial load but when it is mixed with domestic sewage, a microbial infection may occur. Total coliforms and faecal coliforms are the units of measurement.

The gram-negative, facultative, rod-shaped coliform bacteria Escherichia coli can cause serious food contamination due to certain cerotypes [14]. Diseases of the digestive system, urinary tract, newborn meningitis, hemorrhagic colitis, and Cohn's disease can be caused by some of its more dangerous varieties [12, 15]. Severe stomach discomfort, blood in the stool, hemorrhagic colitis, vomiting, and occasionally fever are common signs and symptoms. Bloody diarrhoea, peritonitis, mastitis, septicemia, gram-negative pneumonia, hemolytic-uremic syndrome, and perforation are uncommon complications of virulent strains. Hemolytic uremic syndrome and other serious infections caused by E. coli are more common in younger children, although healthy people of any age can get sick from the infection [15].

References

1. Kannan V, Ramesh R, Sashikumar C (2005) Study on groundwater characteristic and the effects of discharged effluents from textile units at Karur District, India. J Environ Bio 26(2):269–272
2. Kaushal SS, Likens GE, Utz RM, Pace ML, Grese M, Yepsen M (2013) Increased river alkalinization in the eastern U.S. envirornmental science and technology. https://doi.org/10.1021/es401046s
3. U.S. EPA. Report on the 2013 U.S. Environmental protection agency (EPA) international decontamination research and development conference. Research Triangle Park, NC, November 05–07, 2013. U.S. Environmental Protection Agency, Washington, DC, EPA/600/R-14/210, 2014

4. DeZuane J (1997) Handbook of drinking water quality (2nd ed.). John Wiley and Sons. ISBN 0-471-28789-X
5. IS 10500 (2012) Bureau of Indian standards 2012, Specification of drinking water standard. Manak Bhavan, 9 Bahadur Shah Zafar Marg New Delhi 110002
6. Jimbardo PG, Boyd J (1999) Putting Time in perspective: a valid, reliable individual-differences metric. J Pers Soc Psychol 77:1271–1288
7. Navarro RE, Shim EY, Kohara Y, Singson A, Blackwell TK (2001) cgh-1, a conserved predicted RNA helicase required for gametogenesis and protection from physiological germline apoptosis. In: C. elegans, Development, 128(17):3221–32
8. Ferner DJ (2001) Toxicity, heavy metals. eMed. J 2(5):1–7
9. Shelfers SM, Verma S, Cooks RG (1983) Characterization of organic dyes by secondary ion mass spectrometry. Anal Chem 55:2260–2266
10. Aslan G, Akkaya C (2001) Basic problems in groundwater sources and interactions between surface and groundwater. Groundwaters and Environment Symposium, 21–23 March, Izmir, pp 45–54
11. Khadka RB, Khanal AB (2008) Environmental management plan for melamchi water supply project, Nepal. Environ Monitor Assess 146(1–3):225–234
12. Weber EJ, Adams RL (1995) Chemical and sediment mediated reduction of the azo dye Disperse Blue 79. Environ Sci Technol 29:1163–1170
13. Zollinger H (2003) Color chemistry: syntheses, properties, and applications of organic dyes and pigments. Book, John Wiley & Sons
14. Tenaillon O, Skurnik D, Picard B, Denamur E (2010) The population genetics of commensal Escherichia coli. Nat Rev Microbiol 8:207–217. https://doi.org/10.1038/nrmicro2298
15. Lim U, Subar AF, Mouw T, Hartge P, Morton LM, Stolzenberg-Solomon R, Campbell D, Hollenbeck AR, Schatzkin A (2006) Consumption of aspartame-containing beverages and incidence of hematopoietic and brain malignancies. Cancer Epidemiol Biomarkers Prev 15(9):1654–9

Integrated Urban Water Management Plans

Rishika Mukhopadhyay

Many cities are now quite worried about water scarcity. Fast population expansion, insufficient planning, pollution, poverty, and conflicting demands on water resources are the main components of urbanisation that contribute to water stress. Large cities utilise a lot of energy that is often generated elsewhere and import a lot of food and consumer goods that aren't even locally produced. As a result, the impact of urbanization's rising water consumption would be seen much beyond their borders. In developing nations, where most urbanisation is taking place, the necessary water infrastructure for treatment and supply will need substantial planning and financial resources, and the demand for water for direct urban use will also rise. Furthermore, precipitation patterns are predicted to undergo substantial changes due to climate change, impacting water availability and perhaps triggering water-related disasters.

When looking at cost-effectiveness, technical performance, social equality, and environmental sustainability, current paradigms of water management and urban design have either failed or are about to collapse. A change in perspective is necessary across the board. By rethinking water consumption and providing a framework for interventions across the whole water cycle, IUWM planning enables cities to better manage their water resources. Agriculture, industry, households, and ecosystems all consume water, and the IUWM method seeks to strike a balance between their opposing demands.

R. Mukhopadhyay (✉)
Department of Civil Engineering, Elitte College of Engineering, MAKA University of Technology, Kolkata 700113, India
e-mail: mukherjee.rishika@gmail.com

B. G. Gupta (ed.), *Urban Development with Treated Recycled and Reuse of Wastewater: Sustainable Management*, Urban Sustainability,
https://doi.org/10.1007/978-981-96-0857-7_6

1 Characteristics

The following are some of the IUWM concepts that are gaining widespread agreement, and more and more governments are realising the need of implementing them in response to urban challenges:

- For the IUWM plan to be successful, it is essential to involve important actors at all stages: planning, decision-making, implementation, monitoring, and evaluation. To ensure that stakeholders are involved in an organised way, it is necessary to have clearly defined roles and duties under a legal framework.
- To close the "water loop"—that is, to view the whole water cycle as a single entity—it is necessary to have a thorough understanding of the interconnections between water sources, supplies, wastewater, and storm water, and to act accordingly. To better understand urban water systems as a whole, it is helpful to integrate the water cycle's upstream and downstream connections.
- Evaluating a water source portfolio entails exploring possibilities for water sources that have been historically overlooked or unconsidered. Groundwater, precipitation, stormwater, wastewater, and non-sewage wastewater (greywater, for example, water from sinks and dishwashers) all fall into this category.
- Using wastewater to its full potential means actively recovering water, energy, biogas, and nutrients via the application of cutting-edge technology. Improving urban water efficiency relies heavily on recycling and reusing materials.
- Improving resistance to environmental catastrophes and climate change requires the development of adaptive mechanisms as part of the IUWM strategy. Integrated flood control should be a primary focus of IUWM strategies.
- Organisations in the fields of academia, journalism, and activism, as well as businesses and corporations, can work together to form urban water partnerships. Institutional actors must join IUWM and incorporate concepts of supply and demand efficiency into their everyday water consumption, and research and development of new water technologies are equally important.
- Developing world cities in particular need low-cost, high-impact solutions. Urban WASH projects in countries like India, Kenya, and Uganda demonstrate that access to clean water and sanitary facilities does not have to break the bank.

Although there are several obstacles to overcome, there are also numerous opportunities presented by urbanisation, such as improved water management and the provision of sanitation and drinking water to a larger population. Cities can and should be economic engines, places of employment and innovation, and providers of better livelihood opportunities—all while preserving the environment—if IUWM strategies are put into place. Since cities do not stand alone but rather in relation to their hinterlands, it is essential that IWRM plans be in sync with other resource management strategies (e.g. basin, costal, regional, national, etc.).

2 Lessons Learned

There are several lessons to be learned from the cities that have effectively adopted IUWM plans:

- The IUWM approach relies on closing the "water loop" as itself. It is important to view wastewater as a resource and work towards actively recovering and reusing water, energy, biogas, and nutrients from it.
- IUWM plans incorporate an evaluation of all water sources, including black and grey water, among others.
- It is more effective for a single water agency to handle the "water loop" integrally.
- Both raising the availability of water and decreasing the demand for it are necessary for effective and sustainable water management.
- In order to encourage people to use water wisely, public education initiatives are crucial.
- Make water an integral part of cityscapes so that people may experience it firsthand and appreciate its importance.
- Together, we can pave the way for more investment in water technology R&D. Collaborations between the city's commercial, technological, and knowledge-based entities are part of this.

3 Wastewater Treatment and Reuse: A Guide to Help Small Towns Select Appropriate Options

Researchers, economists, and urban developers have been fixated on the meteoric rise of megacities over the last several decades, ignoring the unique problems faced by the world's smaller towns. We can no longer turn a blind eye to the unique infrastructure, service, and housing requirements of small towns, especially when over half of the world's urban population call settlements with less than 500,000 inhabitants home. These municipalities in low- and middle-income countries (LMICs) are seeing significant population growth and are finding it challenging to keep up with the rising demand for wastewater treatment. Safely managed sanitation solutions that are suitable for small towns in terms of size, institutional ability, financial resources, and overall demands are being actively sought after by these communities in an effort to prevent public health problems and subsequent environmental damage.

A new World Bank guide published, **Wastewater Treatment and Reuse: A Guide to Help Small Towns Select Appropriate Options**, delves into the considerations for small town wastewater treatment. Engineers and managers in charge of wastewater service supply may use this guidance as a framework to find and choose the best wastewater treatment methods for their specific kinds of settlements.

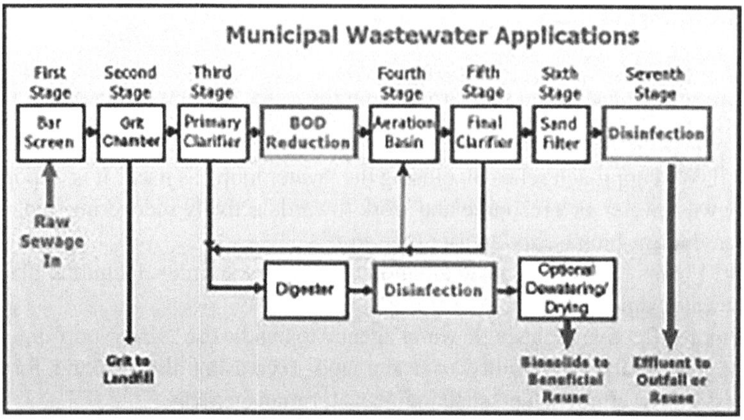

Municipal Wastewater treatment process

Because of their strategic location between urban and rural regions and their importance in wastewater collection and treatment, small towns in LMICs provide a particular challenge. In order for this to occur, it is necessary to choose wastewater treatment solutions that are suitable for small towns. This will help them deal with the challenges of service provision without the advantage of economies of scale and with limited resources for operation and maintenance.

A five-step process for investigating suitable wastewater treatment technologies is proposed in Small-Town Wastewater Treatment, which takes into account the unique circumstances of small towns. The technique has been used to choose and apply various criteria in projects in various areas; case studies in El Salvador, Morocco, and Vietnam illustrate this point. Design of the treatment plant, potential for reuse, and receiving water quality are some of the important decision-making considerations that are addressed. The book also assists users in defining the features and surroundings of a specific town that influence technology choice, drawing on the principles of the Citywide Inclusive Sanitation (CWIS) approach.

Recognising that small towns provide a different reality than both major metropolitan centres and rural locations is the first step in identifying suitable wastewater treatment options for these communities. Managers and engineers in the sanitation industry need to be able to think creatively beyond the box when it comes to technology. They also need to be aware of how local factors like road connections, accessible chemicals and replacement parts, and financial and human resources might impact technological choices.

In our ongoing pursuit of the Sustainable Development Goals, the "no-one left behind" targets continue to pose the greatest threat to small communities. As these communities are located at the crossroads of rural and urban areas, finding effective wastewater treatment solutions for them is crucial to creating a society where everyone has access to clean water.

4 Water-Sensitive Urban Design (WSUD)

To reduce environmental degradation and maximise recreational and aesthetic appeal, WSUD integrates the urban water cycle—including groundwater, storm water, wastewater management, and water supply—into urban design. This technique is based on land planning and engineering [1]. Similar to the American words "low-impact development" (LID) and "Sustainable Drainage System" (SuDS), the Australian and Middle Eastern words "WSUD" are used interchangeably.

Reducing the consumption of potable water and collecting runoff (including wastewater, storm-water, grey-water, and other types of runoff) for recycling purposes are common ways. It is possible to incorporate water filtration, collection, and storage into existing infrastructure.

4.1 Background

A tram running on green tracks in Adelaide, Australia. Replacing impermeable surfaces with grassy surfaces reduces storm water runoff, among other environmental benefits.

Large amounts of storm-water runoff need control because conventional urban and industrial development transforms landscapes from green surfaces that allow water to percolate to a network of impermeable surfaces. Australia, like many other developed nations, has long seen stormwater runoff as an annoyance and potential threat to people's health and property. As a result, there has been an emphasis on designing stormwater management systems that skip ecosystem preservation in favour of swiftly transferring stormwater runoff to streams [2]. This management approach results in what is referred to as urban stream syndrome [3]. Rapid inflow of pollutants and sediments from impermeable surfaces into streams after heavy rains causes streams to have higher concentrations of nutrients, suspended particles, and contaminants. Sedimentation becomes much more prevalent and biotic richness is significantly reduced as peak flow increases, which changes the shape and stability of channels.

Some progress towards comprehensive stormwater management occurred in Australia in the 1960s, when urban stream syndrome was more widely recognised [2]. The federal government and scientists collaborated through the Cooperative Research Centre programme in the 1990s, leading to a significant rise in awareness [4, 5]. The need for an integrated management approach to potable, waste, and stormwater is being more acknowledged by city planners. This will help cities adapt and become more resilient in the face of the challenges posed by climate change, increasing urban densification, and an ageing and costly water infrastructure. Australia is already one of the most drought-stricken countries in the world (2000–2010), and its reliance on surface water sources makes it even more susceptible to climate change. As a result, big cities in the country are facing worsening water shortages [2]. Consequently, stormwater management strategies are evolving as people start to view stormwater runoff as more than just an annoyance and burden [2].

Western Australia was the first Australian state to publish WSUD guidelines in 1994, after the Federal government's 1990s foundational study. In 1999, both Queensland and Victoria issued documents outlining best practices for managing urban stormwater in an environmentally responsible manner. In 1999, Queensland's materials were drafted in conjunction with New South Wales, while Queensland's were distributed through Brisbane City Council [2]. With the goal of improving water efficiency across Australia, the federal, state, and territory governments came together in June 2004 to sign the National Water Initiative (NWI). The National Water Improvement Initiative (NWI) is an all-encompassing plan to better manage water across the nation. It covers a lot of ground and promotes the use of best practices in water management, such as WSUD, across Australia [6].

5 Differences from Conventional Urban Storm Water Management

WSUD sees urban storm water runoff more as an asset than a problem. The approach to water infrastructure and natural resources in urban planning and architecture has undergone a sea change as a result of this [1]. According to WSUD principles, every body of water has many uses and influences on many different things, including biodiversity, water, land, and the aesthetic and recreational value of rivers to the community.

Principles [5]

- Preserving and improving urban waterways include rivers, marshes, and creeks.
- Preserving and enhancing the water quality of runoff from cities and other urban areas into rivers, marshes, and streams.

- Raising the urban water balance via reusing and recycling as much water as possible.
- Reusing water and improving system efficiency can help reduce water use.
- Making stormwater treatment a part of the environment in a way that improves water quality while also providing habitat for animals, recreational opportunities, and open public space.
- Reducing urban runoff and peak flows while increasing infiltration and recharging groundwater.
- Adding aesthetic, social, cultural, and ecological benefits to urban planning through water integration.
- The broad adoption of WSUD is made possible by its easy and cost-effective setup.

Objectives [1]

- Reducing the demand for potable water by managing water on both the demand and supply sides.
- Using fixtures and appliances that conserve water.
- Using rainwater and other possible alternative water sources in a purpose-fitting manner.
- Reducing wastewater production while treating it to an acceptable level for effluent reuse or discharge to recipient waters.
- Reusing or discharging treated stormwater in a way that satisfies water quality standards involves removing sediments, contaminants, and nutrients by retaining the water and releasing it slowly.
- Enhancing the condition of waterways by using treatment and reuse technologies to return or maintain the catchments' natural hydrological regime.
- Enhancing the aesthetic value of urban residents' relationship with water.
- By incorporating localised water storage into urban planning, we may maximise the use of water sources to reduce the input and outflow of potable, storm, and waste water, thereby promoting a high level of water-related self-sufficiency within urban contexts.
- Water and plants can help mitigate the "urban heat island effect" and restore depleted groundwater supplies.

Techniques [1]

- Utilising water-efficient equipment to decrease use of potable water.
- Using grey water for reuse instead of wasting it helps reduce the demand for fresh water.
- Stormwater collection, rather than quick transportation, of stormwater.
- Avoiding the need to increase drainage systems by reuse, storage, and infiltration of stormwater.
- Utilisation of plants for the purpose of stormwater filtration.
- Reducing potable water usage with water-efficient landscaping.

- The preservation of cultural, recreational, and environmental values connected to water through the reduction of a project's ecological impact on the provision of supply, wastewater, and stormwater services.
- Reuse and localised wastewater treatment systems to lessen the need for potable water and lessen the negative impact on the environment from wastewater discharges.
- Meeting the water needs of altered waterways by the use of rainwater or other recycled urban waters, with the necessary restrictions in place in every instance.
- Institutional frameworks that are adaptable enough to deal with climate change's growing unpredictability and variability.
- An emphasis on preparing for the future, including in ancillary fields like water demand management.
- A combination of centralised and dispersed water infrastructure, backed by a varied portfolio of water sources.

Common WSUD Practices

Listed below are some of the most common WSUD practices in Australia. To achieve goals in urban water cycle management, a mix of these is often used.

6 Road Layout and Street Scape

6.1 Bio-retention Systems

Filtration of silt and other particles via specified medium follows the treatment of water by plants in bioretention systems. Some pollutants, such as soluble or fine-particle nitrogen and phosphorus, may be biologically uptaken by vegetation. A popular method of filtering and treating runoff before it reaches street drains, bioretention systems have a smaller footprint than alternatives like built wetlands. Other devices could be more suited for use on bigger sizes because of the complexity involved. Bioretention basins and bioretention swales are the two main components of a bioretention system.

6.2 Bio-retention Swales

The median strip of split roadways is a common location for bioretention swales, which are similar to buffer strips and swales but are built into the base of a swale. Stormwater treatment and are both handled by them. Treatment needs determine whether a bioretention system is implemented over the whole length of a swale or just in sections. Typically, runoff water is filtered using fine media before being sent downwards via a perforated pipe to be collected and sent to storage areas or rivers further downstream. As an alternative to infiltration systems, bioretention swales may be used with a variety of soil types and can avoid erosion by allowing vegetation to grow in the filter medium [7].

6.3 Bio-retention Basins

Parking lot that drains to a small bio-retention basin

Bio-retention swales and bio-retention basins serve comparable purposes in controlling flow and treating water quality, although bioretention basins do not carry water [7]. To make the most of runoff treatment during low to medium flow events, basins not only filter the water and absorb biological materials, but they also detain rainwater for a longer period of time. Although it is more often used to represent smaller, individual lot-scale bioretention basins, the term "raingarden" may also describe such systems [1]. Bioretention basins may be strategically placed inside projects because to their adaptability to various sizes and forms. They are placed at regular intervals along roadways, similar to other bioretention systems, to treat runoff before it enters the drainage system [7]. At the drainage system's outfalls, for example, a bigger basin can treat a wider region. Within a bioretention basin, you may grow a variety of plants that blend in well with the rest of your landscaping. It is recommended to choose vegetation species that can withstand recurrent flooding [1]. Bioretention basins, on the other hand, are quite vulnerable to substances that could obstruct the filter medium. To minimise harm to plant or filter media surfaces, basins are often used in combination with litter traps, also known as gross pollutant traps (GPTs), and coarser sediment basins. The latter two types of basins collect litter and other large particles, while the former includes the ubiquitous trash racks.

6.4 Infiltration Trenches and Systems

A underground reservoir may be formed by filling infiltration trenches, which are shallowly dug structures, with permeable materials like gravel or rock [7]. Their primary function is to collect rainwater runoff in a subterranean trench, which is then slowly released into the soil and groundwater systems in the area [1]. Even though they aren't usually intended to cure anything, they may help with that by holding onto contaminants and sediments. Flows may be captured and infiltrated to lower runoff volumes and peak discharges from impermeable surfaces.

Infiltration systems are often placed as the last component of a WSUD system since their main role is to release treated storm-water [7]. Do not put infiltration trenches on very unstable or steep hillsides. It is common practice to line the trench with geotextile cloth to stop soil erosion and soil migration into rock or gravel fill. Depending on the local soil characteristics, infiltration systems work best with sandy-loam soils that have deep groundwater and a high infiltrative capacity. It is possible to embed a perforated pipe in gravel in locations with low-permeability soils, such as clay.

To keep the system clear of sediments and at the correct infiltration rate, regular maintenance is essential. Checking and maintaining the pre-treatment involves inspecting it periodically and clearing out any blocked material.

6.5 Sand Filters

Sand filters function similarly to bio-retention systems and are an adaptation of the infiltration trench concept. They purify stormwater before it is released into the stormwater system further downstream. To treat runoff from densely populated and urbanised regions, as well as from enclosed hard surfaces like parking lots, sand filters are an excellent choice [1]. In most cases, they cannot sustain plant life due of their subsurface installation and the fact that the filtering medium, sand, does not keep enough moisture. In most cases, the filter will include a sedimentation chamber that acts as a pre-treatment device, removing larger sediments, debris, and pollutants; a weir to further sort the sediments; and finally, a sand layer to filter out smaller particles, dissolved pollutants, and sediments. Like bioretention systems, the filtered water is collected by means of perforated underdrain pipes [7]. A system's overflow chamber is another possible component. In the case of a storm, the sedimentation chamber might be built to drain using weep holes or to contain permanent water. Anaerobic conditions, which may result in the discharge of contaminants like phosphorus, pose a danger to permanent water storage. Consideration of detention storage for high hydrologic efficacy and discharge management via appropriate size of the perforated underdrain and overflow channel should be part of the design process. Preventing crust formation requires regular maintenance.

6.6 Porous Paving

Porous paving (or pervious paving) is an alternative to conventional impermeable pavement and allows infiltration of runoff water to the soil or to a dedicated water storage reservoir below it [7, 8]. It minimises the amount and velocity of stormwater runoff and may enhance water quality by eliminating pollutants via filtration, interception, and biological treatment in generally flat places like car parks, driveways, and rarely used roadways [9]. There are a variety of monolithic and modular porous pavement designs to choose from. In contrast to modular buildings that include individual pavers with spaces between them, monolithic structures are made up of a single continuous porous material like asphalt or porous concrete [7]. Pavements manufactured from minimally-materialized asphalt or concrete, concrete grid pavements, and concrete, ceramic, or plastic modular pavements are some examples of commercially accessible goods [1]. A layer of geotextile material is often deposited under a highly porous material, such as sand or gravel, to create a porous pavement. The diverse types of porous pavement call for different maintenance procedures. It is recommended to remove dirt and debris and conduct inspections. In the event that blockages do arise, the affected pavers may be raised, backwashed, and replaced [7]. Generally porous pavement is not suited for areas with heavy traffic loads [9]. Stormwater particles have the potential to block material pores.

7 Public Open Space

7.1 Sedimentation Basins

Sediment basin installed on a construction site

Sedimentation basins (also called sediment basins) are often the first component of a wastewater treatment unit design for WSUD because of its dual purpose: removing coarse to medium-sized sediments by settling them and regulating water flows [7]. They help sediments settle out of the water by temporarily retaining rainwater and reducing flow velocities. In order to prevent downstream components from being overwhelmed or choked by coarse sediments, they play a crucial role as a pre-treatment. Incorporating sedimentation basins into an urban design as a permanent system or using them as a stopgap solution during construction are two examples of their many potential applications. Inlet ponds to bio-retention basins or man-made wetlands are common uses for these types of ponds. Most sedimentation basins are intended to remove 70 to 90% of coarser particles (125 μm and greater), and they are usually the most successful at this task [1]. They may be made to either have a permanent pool or to drain during dry spells and fill up during rainstorms. A secondary spillway diverts water away from the basin and into a bypass channel or conveyance system when the flow exceeds the intended discharge, keeping the sediments there from being re-suspended.

7.2 *Constructed Wetlands*

Constructed wetlands are created with the express purpose of removing dissolved toxins and fine to colloidal particles from rainwater. To rid themselves of these contaminants, these shallow, densely vegetated bodies of water use improved sedimentation, fine filtration, and biological absorption [7]. Typically, these systems are divided into three zones: an intake zone, also known as a sedimentation basin, which is used to remove larger sediments; a macrophyte zone, which is thickly vegetated and is responsible for removing smaller particles and soluble pollutants; and finally, a high flow bypass channel, which is used to protect the macrophyte zone [1]. The macrophyte zone often consists of both open water and marsh areas, is between 0.25 and 0.5 m deep, is home to specialised plant life, and has a retention period of 48–72 h. An additional role that constructed wetlands may play in flow management is to rise during rainstorms and gradually release the stored flows [10]. Depending on the wetland processes, constructed wetlands may increase the quality of runoff water. Wetlands primarily trap suspended solids and adsorbed pollutants through physical means; they also trap dissolved pollutants through biological and chemical uptake; and they transform pollutants through more stable sediment fixation, microbiological processes, and ultraviolet disinfection [10].

To prevent typical issues like weed infestation, mosquito difficulties, algal blooms, and the buildup of garbage, oil, and scum in certain areas of the wetland, manmade wetlands must be meticulously planned [7]. A lot of space has to be set aside for constructed wetlands, and they can't be built on hills. Construction wetlands might be a disincentive to WSUD measures due to the high price of the area and vegetation growth [7]. Guidelines for developers (such as the Urban Stormwater: Best Practice Environmental Management Guidelines in Victoria [11]) must be included into the design in order to effectively trap and hold particles measuring 125 μm or below, while simultaneously decreasing common pollutants like nitrogen and phosphorus by a minimum of 45%. Enhanced aesthetic and recreational qualities, habitat supply, and stormwater management are all part of the design requirements for manmade wetlands [10]. To maintain a synthetic wetland in good repair, it is common practice to remove trash and silt from the inlet zone, manage weeds and harvest macrophytes every so often to ensure a healthy ecosystem [7].

7.3 Swales and Buffer Strips

Two swales for a housing development. Both the foreground and backdrop ones are currently at various stages of development

Swales and buffer strips serve as an alternative to pipes for transporting stormwater and provide a protective barrier between the impervious parts of a catchment region and the receiving waters (such as a stream or a wetland). Water moves more evenly downstream when it encounters gentle slopes and overland flows. Treatment occurs in buffer zones as a result of sedimentation and interactions with plants.

Urban designers may enhance the visual character of a region by incorporating swales along roadways or parklands. Commonly, swales are designed with longitudinal slopes ranging from one percent to four percent. This helps to keep the flow capacity high while preventing hazards like erosion of the bioretention or swale surface and high velocities [1]. To decrease speed and disperse flows equally across swales, thick vegetation or check banks along swales might be used in steeper locations [7]. Underdrains are a potential solution for water-logging and stagnant ponding in gently sloping swales. The vegetation that will be planted in the swale has to be dense enough to filter water effectively and strong enough to resist the design flows [7]. The ideal height for vegetation is to be higher than the water levels in the treatment flow. When water flows into a swale in a straight line, against the direction of the main flow, the swale's edge filters out the runoff and functions as a buffer.

7.4 Ponds and Lakes

An artificial body of open water, such as a pond or lake, may be formed by building a dam wall and installing a weir outlet structure [7]. They can treat runoff in the same way as man-made wetlands do by extending detention time and facilitating sedimentation, nitrogen absorption, and UV disinfection. They are also aesthetically pleasing, offer a home for animals, and store water that might be utilised for irrigation or other purposes down the road [12]. Often, artificial ponds and lakes also form part of a flood detention system [1]. In man-made bodies of water, such as lakes and ponds, aquatic plants are crucial for controlling nutrient and oxygen levels and preserving water quality. Submerged plants may be found in the open water zone, however emergent macrophytes are often limited to the edges when the water depth is more than 1.5 m. Vegetation that forms a border around a bank may help slow its erosion. As a pretreatment, ponds are often utilised in conjunction with sediment basins or artificial wetlands rather than as a standalone WSUD technique.

Despite their aesthetic value, many lakes and ponds are unhealthy due to factors such as insufficient inflows to maintain lake levels, insufficient flushing of the lake due to excessive residence time, improper mixing (stratification) that results in low dissolved oxygen levels, and/or poor water quality of inflows [12]. Blue-green algae endangers lake health when combined with low water quality and excessive fertiliser levels. Consideration of catchment hydrology and water level, as well as the architecture of the pond or lake (directed to prevailing winds to enable mixing), are crucial factors in the construction of lakes and ponds to guarantee their long-term sustainability. Proper pre-treatment and the avoidance of significant nutrient "spikes" should be the goals of the design of hydraulic structures (inlet and outflow zones). Designing the landscape with the right plant species and density of planting is also essential [7]. A lot of people aren't interested in using ponds and lakes as WSUD measures because of the high expenditures of the planned area, planting vegetation, and the regular upkeep that's needed.

The best way to keep pond and lake systems healthy is to keep them well-maintained. It is common practice to remove plants, detritus, weeds and litter from the intake zone and to replant every so often. It may be required, in some instances, to artificially flip the lake.

References

1. BMT WBM (2009) Evaluating options for water sensitive urban design—a national guide: prepared by the Joint Steering Committee for Water Sensitive Cities: in delivering Clause 92(ii) of the National Water Initiative. Joint Steering Committee for Water Sensitive Cities (JSCWSC), Canberra, viewed 18 Sep 2011 [1]. Archived 2 June 2011, at the Wayback Machine
2. Roy AH, Wenger SJ, Fletcher TD, Walsh CJ, Ladson AR, Shuster WD, Thurston HW, Brown RR (2008) Impediments and solutions to sustainable, watershed-scale urban stormwater management: lessons from Australia and the United States. Environ Manage 42(2):344–359.

Bibcode: 2008EnMan.42.344R. PMID: 18446406. S2CID7685925. https://doi.org/10.1007/s00267-008-9119-1
3. Walsh CJ, Roy AH, Feminella JW, Cottingham PD, Groffman PM, Morgan RP (2005) The urban stream syndrome: current knowledge and the search for a cure. J North Am Bentholog Soc 24(3):706–723. S2CID 30667397. https://doi.org/10.1899/04-028.1
4. Brown R, Clarke J (2007) The transition toward water sensitive urban design: the story of Melbourne, Australia, Report No. 07/01. Facility for Advancing Water Biofiltration, Monash University, Clayton, VIC
5. Donofrio J, Kuhn Y, McWalter K, Winsor M (2009) Research Article: Water-sensitive urban design: an emerging model in sustainable design and comprehensive water-cycle management. Environ Pract 11(3):179–189. S2CID 131103400. https://doi.org/10.1017/S1466046609990263
6. Council of Australian Governments (2009) Intergovernmental agreement on a national water initiative. Council of Australian Governments, Canberra, viewed 18 Sep 2011 [2]. Archived 2 June 2011, at the Wayback Machine
7. WSUD engineering procedures: stormwater (2008). CSIRO Publishing, Melbourne
8. Fletcher TD, Duncan HP, Poelsma P, Lloyd S (2003) Stormwater flow and quality and the effectiveness of non-proprietary stormwater treatment measures: a review and gap analysis. The Institute for Sustainable Water Resources, Monash University and the CRC for Catchment Hydrology, Victoria
9. Jayasuriya LNN, Kadurupokune N, Othman M, Jesse K (2007) Contributing to the sustainable use of stormwater: the role of pervious pavements. Water Sci Technol 56(12):69–75
10. Melbourne Water (2010) Constructed wetlands guidelines. Melbourne Water, Melbourne
11. Victorian Stormwater Committee (1999) Urban stormwater: best-practice environmental management guidelines. CSIRO Publishing, Collingwood, VIC, viewed 19 Sep 2011
12. Mackay City Council (2008) Engineering design guidelines: constructed lakes. Planning Scheme Policy No. 15.15, Mackay City Council, Mackay, QLD

Water Re-use

Aftabuz Zaman

1 Rainwater Tanks

Rainwater tanks (also known as "rainwater harvesting") are systems that collect precipitation and use it to partly satisfy household water needs, such as in times of drought. This helps to save potable water and alleviates environmental stress. Rainwater tanks may also lessen the amount of storm water runoff and contaminants that make it to downstream streams [1]. They can be used effectively in domestic households as a potential WSUD element [2]. Buildings may collect and use rain and runoff for specialised uses, such flushing toilets, washing clothes, watering gardens, and washing cars [3]. Rainwater collected from hard surfaces may soak into the site via buffer tanks, which helps keep groundwater and aquifer levels stable [4].

No rules or standards in Australia specify quantifiable performance goals for rainwater tanks, such as minimum tank size or reductions in potable water use [1]. However, according to the various state government rules, rain water tanks should be built to be a dependable source of water that may augment the mains water supply while still maintaining the water's proper quality [1]. Supply and demand, water quality, stormwater advantages (reduced volume), cost, available space, maintenance, tank size, form, and material should all be considered when using rainwater tanks. Plumbing and drainage codes must also be followed while installing rainwater tanks [5]. A water filter or first flush diversion, a maintenance drain, a pump (pressure system), a dual supply system, and an on-site retention provision are all components that might be part of an ideal arrangement [1].

Air pollution, possum and bird droppings, insects (such as mosquito larvae), roofing materials, paints, and detergents are all potential contaminants of water

A. Zaman (✉)
School of Agricultural Sciences, Sister Nivedita University, Kolkata, India
e-mail: profazaman@gmail.com

B. G. Gupta (ed.), *Urban Development with Treated Recycled and Reuse of Wastewater: Sustainable Management*, Urban Sustainability,
https://doi.org/10.1007/978-981-96-0857-7_7

quality. Annual flushing (to remove built-up sludge and debris) and visual examinations should be carried out as part of maintenance [1, 6].

2 Aquifer Storage and Recovery (ASR)

Aquifer storage and recovery (ASR) (also referred to as Managed Aquifer Recharge) encourages the use of gravity feed or pumps to replenish water in subterranean aquifers. With the ability to pump water up from below during dry seasons, it can serve as a substitute for massive surface storages [7]. Storm water or treated wastewater are two possible water sources for an ASR system. The following are common parts of an ASR system that collects rainwater [8]:

1. A structure that redirects a drain or stream
2. An apparatus for purifying both the storm water that is collected and the water that is treated before injection
3. A marsh, retention pond, dam, or tank used for short-term storage
4. An overflow or spillage structure
5. Both a water injection well and a water recovery well
6. Devices (such as sample ports) for tracking the quantity and quality of water.

Fragmented unconfined rock and confined sand and gravel are two examples of aquifer types that might work with an ASR system. The viability of an ASR plan can only be determined by comprehensive geological studies. One possible selling point of ASR is its lower cost as compared to underground storage. The preservation of groundwater quality and the appropriate use of recovered water should be considered throughout the design phase. It is also important to prevent depletion and excessive pressures from harming aquifers and aquitards. The locations downstream of the harvesting site should also be considered for their potential impacts. Selecting an aquifer, treating it, injecting it, recovering from it, and then keeping an eye on it all take meticulous preparation.

3 Policy, Planning, and Legislation

Urban water cycle management is not mandated by national law in Australia because of the constitutional separation of powers between the federal government and the individual states. In 2004 and 2006, the federal, state, and territory administrations reached an agreement on the National Water Initiative (NWI), which aims to enhance water management nationwide [9]. It urges the use of WSUD strategies and makes its goal to "Create Water Sensitive Australian Cities" very apparent. To further assist with the assessment of WSUD programmes, national standards have been published in compliance with NWI clause 92(ii) [7].

State planning and environmental laws generally support ESD but have varied degrees of vagueness when it comes to WSUD standards. In some cases, state planning rules lay out more detailed requirements for the implementation of WSUD procedures.

Regional and/or local catchment-scale integrated water cycle management plans and/or storm water management plans provide the strategic framework for WSUD at the local government level [10]. In order to execute WSUD, local government environment plans may impose regulatory restrictions on projects.

Inconsistent WSUD policy and practice implementation and fragmented management of bigger watersheds are consequences of the fact that regulatory power over storm water runoff is divided between the Australian states and local government regions. In Melbourne, for instance, smaller watersheds are governed by municipal governments, but watersheds larger than 60 ha are under the control of Melbourne Water, the state-level body. Because of this, Melbourne Water has been reluctant to spend much on watershed storm water drainage (WSUD) projects that would benefit smaller watersheds, even though they have an impact on the larger watersheds from which they drain and the quality of the waterways, especially the headwater streams.

4 State Legislation and Policy

4.1 Victoria

Many of the overarching goals and tactics of Victoria's planning policy use WSUD principles [11]. A few explicit provisions mandating the implementation of WSUD practices are included in the State Planning Policy Framework of the [Victoria Planning Provisions], which is included in all planning schemes in Victoria [12].

Permeability standards for new residential developments mandate that at least 20 percent of the plots not be covered by impermeable surfaces [11]. This will help to enhance on-site storm-water infiltration and lessen the negative effects of increasing storm-water runoff on the drainage system.

In order for new residential subdivisions consisting of two or more lots to fulfil integrated water management goals, they must:

- drinking water supply [13];
- reused and recycled water [14];
- waste water management [15], and
- urban run-off management [16].

According to Clause 25 of the "Victoria Planning Provisions" (c. 56.07-4) pertaining to urban runoff management, storm water systems are required to achieve best practice goals in storm water management. Even though it isn't considered to be best practice anymore, Urban Storm water: Best Practice Environmental Management Guidelines is still the state standard [17, 18]. Current water quality goals do not shield streams from storm water damage and include:

- 80% of the yearly suspended solids load in a typical metropolitan area is retained.
- Retention of the average yearly total phosphorus load in an urban setting is 45%.
- Retention of the average urban yearly total nitrogen load is 45%.
- 70% less litter each year in an average metropolitan area.

Additionally, the needs of the appropriate drainage authority must be satisfied by urban storm water management systems. It is common for this to be the local council [19]. On the other hand, Melbourne Water is in charge in the Melbourne area for catchments larger than 60 hectares. Unless the appropriate drainage authority gives its approval and there are no negative repercussions downstream, inflows into the watershed beyond the subdivision site must remain at pre-development levels.

An online tool called SSTORM (Storm water Treatment Objective—Relative Measure) is provided by Melbourne Water. It helps users determine whether development projects match the storm water quality performance criteria set by legislation. Given that the STORM tool can only evaluate individual WSUD treatment procedures, it cannot account for situations in which a combination of treatments is used [20]. Additionally, it may only be used on locations with more than 40% covered by impermeable surfaces. Highly advanced modelling tools, like MUSIC software, are suggested for bigger and more intricate projects.

4.2 New South Wales

In New South Wales, the main policy that requires WSUD implementation is the State Environmental Planning Policy (Building Sustainability Index: BASIX) 2004 (NSW). With BASIX, a user may input information about a residential complex—its location, size, construction materials, etc.—and then obtain a score based on how well the property meets water and energy efficiency goals. Residential developments have water reduction objectives that vary in intensity and location, ranging from zero to forty percent reduction in mains-supplied potable water usage (for more on this, see water demand management) [21]. The 40% water goal is met for 90% of newer dwellings. Greywater recycling, rainwater tanks, and stormwater tanks are a few of the WSUD components that may be modelled with the help of the BASIX project.

According to the Local Government Act of 1993 (NSW), it is the responsibility of local councils to create Local Environment Plans (LEPs). These plans have the power to regulate development and require the implementation of WSUD practices and goals. However, local council acceptance is inconsistent due to a lack of state-level policy and guidance; some have included water, sanitation, and urban development (WSUD) goals in their LEPs, while others do not [22].

In 2006 the then NSW Department of Environment and Conservation released a guidance document, Managing Urban Stormwater: Harvesting and Reuse.

The document presented an overview of stormwater harvesting and provided guidance on planning and design aspects of integrated landscape-scale strategy as well as technical WSUD practice implementation [23]. Even though the paper is still up on the government website, it doesn't seem to be getting any attention.

In addition, local councils may access materials and tools from the Sydney Metropolitan Catchment Management Authority to help them implement WSUD [24]. These include:

- With the permission of state-level departments in New South Wales, possible WSUD requirements might be included into Local Government LEPs [25];
- Possible WSUD provisions for inclusion in documents prepared for or submitted to the Local Government [26];
- The WSUD Decision Support Tool, which helps local governments compare and assess WSUD programmes in the field [27], and
- Outline of procedures for the usage of the more advanced MUSIC modelling software in New South Wales [28].

References

1. WSUD engineering procedures: stormwater (2008). CSIRO Publishing, Melbourne
2. Khastagir A, Jayasuriya LNN (2010) Impacts of using rainwater tanks on stormwater harvesting and runoff quality. Water Sci Technol 62:324–329
3. "Managing stormwater on your property. City of Albany, Western Australia. Archived from the original on 21 Apr 2013. Retrieved 10 Nov 2012
4. Final report, Water Smart Australia Project, Restoration of the groundwater aquifer on the Cottlesoe Peninsula, Western Australia (PDF). Archived from the original (PDF) on 22 Mar 2012. Retrieved 10 Nov 2012
5. Standards Australia (2003) AS/NZS 3500.3: plumbing and drainage—stormwater drainage. Standards Australia, Sydney
6. Abbott SE, Douwes J, Caughley BP (2006) A survey of the microbiological quality of roof-collected rainwater of private dwellings in New Zealand. NZ J Environ Health 29:6–16
7. BMT WBM (2009) Evaluating options for water sensitive urban design—a national guide: prepared by the Joint Steering Committee for Water Sensitive Cities: in delivering Clause 92(ii) of the National Water Initiative. Joint Steering Committee for Water Sensitive Cities (JSCWSC), Canberra, viewed 18 Sep 2011 [1]. Archived 2 June 2011, at the Wayback Machine
8. Environmental Protection Agency (South Australia) (2004) Code of practice for aquifer storage and recovery, Adelaide, viewed 19 Sep 2011 [3]. Archived 12 Oct 2009, at the Wayback Machine
9. Council of Australian Governments (2009) Intergovernmental agreement on a national water initiative. Council of Australian Governments, Canberra, viewed 18 Sep 2011, [2]. Archived 2 June 2011, at the Wayback Machine
10. NMcAuley A, McManus R, Knights D (2009) Water sensitive urban design: implementation framework for Darwin discussion paper. Northern Territory Government, Sydney, viewed 18 Sep 2011. Archived 22 Mar 2011, at the Wayback Machine
11. Victoria Planning Provisions cs. 14.02-1, 14.02-3, 19.03-2, 19.03-3
12. Victoria Planning Schemes
13. Victoria Planning Provisions c. 56.07-1. Archived 20 Sep 2011, at the Wayback Machine
14. Victoria Planning Provisions c. 56.07-2. Archived 20 Sep 2011, at the Wayback Machine

15. Victoria Planning Provisions c. 56.07-3. Archived 20 Sep 2011, at the Wayback Machine
16. Victoria Planning Provisions c. 56.07-4. Archived 20 Sep 2011, at the Wayback Machine
17. Victorian Stormwater Committee (1999) Urban stormwater: best-practice environmental management guidelines. CSIRO Publishing, Collingwood, VIC, viewed 19 Sep 2011
18. Victoria Planning Provisions c. 56.07. Archived 20 Sep 2011, at the Wayback Machine
19. State of Victoria, Department of Sustainability and Environment (2006) Using the integrated water management provisions of Clause 56—residential subdivision. State of Victoria, Department of Sustainability and Environment, Mulgrave, VIC, viewed 19 Sep 2011. http://www.dpcd.vic.gov.au/__data/assets/pdf_file/0020/41717/VPP_Clause_56_4-Intwaterman.pdf. Archived 11 Sep 2011 at the Wayback Machine
20. Melbourne Water (n.d.) About STORM, viewed 19 Sep 2011. https://storm.melbournewater.com.au/help/about_storm.asp
21. BASIX (2006) BASIX: about the BASIX indices, viewed 19 Sep 2011, [5]. Archived 22 Apr 2011, at the Wayback Machine
22. McManus R, Morison P (2007) Barriers and opportunities to WSUD adoption in the Botany Bay catchment, viewed 19 Sep 2011
23. Department of Environment and Conservation NSW (2006) Managing urban stormwater: harvesting and reuse. Department of Environment and Conservation NSW 2006, Sydney, viewed 19 Sep 2011
24. Sydney Metropolitan Catchment Management Authority (2010) Tools & resources, viewed 19 Sep 2011
25. WSUD Clauses (PDF). Archived from the original (PDF) on 6 Apr 2011. Retrieved 21 Sep 2011
26. Tools & Resources. WSUD 17 May 2012. Archived from the original on 05 May 2012. Retrieved 21 Jan 2014
27. Draft NSW MUSIC Modelling Guidelines (PDF), p 81. Archived from the original (PDF) on 6 Apr 2011. Retrieved 11 Mar 2023
28. BMT WBM (2010) Draft NSW MUSIC modelling guidelines, Sydney, viewed 17 Sep 2011 [6]. Archived 6 Apr 2011, at the Wayback Machine

Predictive Modelling to Assess WSUD Performance

Biman Gati Gupta

In order to evaluate if WSUD practices are being implemented in accordance with local rules, several countries provide simplified modelling programmes.

STORM is supplied by Melbourne Water, whereas BASIX is used in residential constructions in New South Wales, Australia. Bigger and more complex projects could call for better modelling tools [1].

1 Issues Affecting Decision-Making in WSUD

1.1 Impediments to the Adoption of WSUD

Major issues affecting the adoption of WSUD include [2]:

"Regulatory framework barriers" and institutional fragmentation at state and local government levels.
"Assessment and costing" uncertainties relating to selecting and optimising WSUD practices for quantity and quality control.
"Technology and design" and complexity integrating into landscape-scale water management systems.
"Marketing and acceptance" and related uncertainties.

The transition of Melbourne city to WSUD over the last forty years has culminated in a list of best practice qualities [3] and enabling factors [4], which have been identified as important in aiding decision making to facilitate transition to WSUD

B. G. Gupta (✉)
Elitte College of Engineering, MAKA University of Technology Sodepur, Kolkata 700113, India
e-mail: bimangupta@yahoo.in

B. G. Gupta (ed.), *Urban Development with Treated Recycled and Reuse of Wastewater: Sustainable Management*, Urban Sustainability,
https://doi.org/10.1007/978-981-96-0857-7_8

technologies. The implementation of WSUD can be enabled through the effective interplay between the two variables discussed below [5].

2 Qualities of Decision-Makers

- **Vision for Waterway Health**—A common vision for waterway health through cooperative approaches.
- **Multi-sectoral Network**—A network of champions interacting between government, academia, and private sector.
- **Environmental Values**—Strong environmental protection values.
- **Public-Good Disposition**—Advocacy and protection of the public good.
- **Best-Practice Ideology**—Pragmatic approach to aid cross-sectoral implementation of best practices.
- **Learning-by-Doing Philosophy**—Adaptive approach to incorporating new scientific information.
- **Opportunistic**—Strategic and forward thinking approach to advocacy and practice.
- **Innovative and Adaptive**—Challenge status quo through focus on adaptive management philosophy.

3 Key Factors for Enabling WSUD

- "Socio-political Capital"—An aligned community, media, and political concern for improved waterway health, amenity, and recreation.
- "Bridging Organization"—Dedicated organizing entity that facilitates collaboration between science and policy, agencies and professions, and knowledge brokers and industry.
- "Trusted and Reliable Science"—Accessible scientific expertise, innovating reliable and effective solutions to local problems.
- "Binding Targets"—A measurable and effective target that binds the change activity of scientists, policy makers, and developers.
- "Accountability"—A formal organizational responsibility for the improvement of waterway health, and a cultural commitment to proactively influence practices that lead to such an outcome.
- "Strategic Funding"—Additional resources, including external funding injection points, directed to the change effort.
- "Demonstration Projects and Training"—Accessible and reliable demonstration of new thinking and technologies in practice, accompanied by knowledge diffusion initiatives.
- "Market Receptivity"—A well-articulated business case for the change activity.

4 WSUD Projects in Australia

WSUD technologies can be implemented in a range of projects, from previously pristine and undeveloped, or green field sites, to developed or polluted brown field sites that require alteration or remediation. In Australia, WSUD technologies have been implemented in a broad range of projects, including from small-scale road-side projects, up to large-scale + 100 hectare residential development sites. The three key case studies below represent a range of WSUD projects from around Australia.

References

1. Cooperative Research Council for Catchment Hydrology (2000) Water sensitive urban design in the Australian context: synthesis of a conference held 30–31 August 2000, Melbourne, Australia. Cooperative Research Council for Catchment Hydrology, Melbourne, viewed 19 September 2011
2. Brown R, Clarke J (2007) The transition toward water sensitive urban design: the story of Melbourne, Australia. Report No. 07/01, Facility for Advancing Water Biofiltration, Monash University, Clayton, VIC, p 44
3. Brown R, Clarke J (2007) The transition toward water sensitive urban design: the story of Melbourne, Australia. Report No. 07/01, Facility for Advancing Water Biofiltration, Monash University, Clayton, VIC, p 47
4. Brown R, Clarke J (2007) The transition toward water sensitive urban design: the story of Melbourne, Australia. Report No. 07/01, Facility for Advancing Water Biofiltration, Monash University, Clayton, VIC, p V
5. WSUD (2010). WSUD roadway retrofits, Sydney, viewed 19 Sep 2011 [dead link]

A Rain Garden Bio Filter for Small-Scale Storm Water Management

Aftabuz Zaman

1 Ku-ring-gai Council's Kooloona Crescent Rain Garden, NSW

The WSUD Roadway Retrofit Bioretention System is a small-scale project implemented by the Ku-ring-gai Council in NSW as part of an overall catchment incentive to reduce stormwater pollution. The Raingarden uses a bioretention system to capture and treat an estimated 75 kg of total suspended solids (TSS) per year of local stormwater runoff from the road, and filters it through a sand filter media before releasing it back into the stormwater system. Permeable pavers are also used in the system within the surrounding pedestrian footpaths, to support the infiltration of runoff into the ground water system [1]. Roadside bioretention systems similar to this project have been implemented throughout Australia. Similar projects are presented on the Sydney Catchment Management Authority's WSUD website [2]:

- 2005 Ku-ring-gai Council—Minnamurra Avenue Water Sensitive Road project [3];
- 2003 City of Yarra, Victoria—Roadway reconstruction with inclusion of bioretention basins to treat stormwater [4];
- 2003–2004 City of Kingston, Victoria (Chelsea)—Roadway reconstruction with inclusion of bioretention basins to treat stormwater [5], and
- 2004 City of Kingston, Victoria (Mentone)—Roadway reconstruction with inclusion of bioretention basins to treat stormwater [6].

A. Zaman (✉)
School of Agricultural Sciences, Sister Nivedita University, Kolkata, India
e-mail: profazaman@gmail.com

B. G. Gupta (ed.), *Urban Development with Treated Recycled and Reuse of Wastewater: Sustainable Management*, Urban Sustainability,
https://doi.org/10.1007/978-981-96-0857-7_9

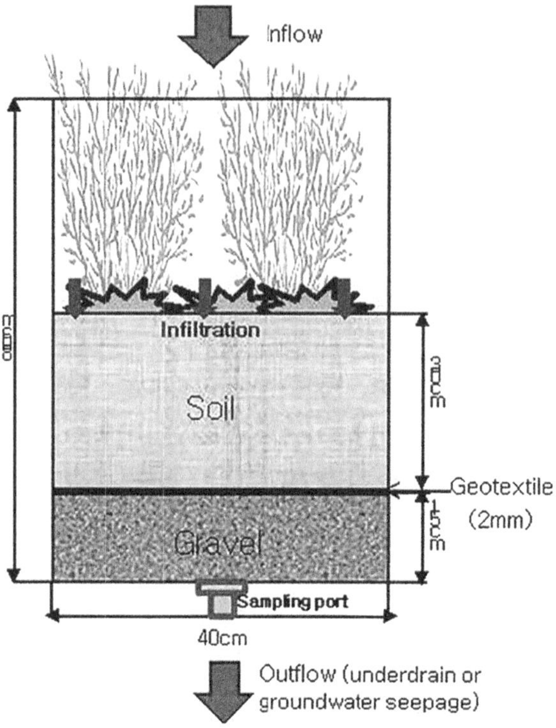

Sketch of rain garden

2 WSUD in Residential Development Projects

2.1 Lynbrook Estate, Victoria

The Lynbrook Estate development project in Victoria, demonstrates effective implementation of WSUD by the private sector. It is a Greenfield residential development site that has focused its marketing for potential residents on innovative use of stormwater management technologies, following a pilot study by Melbourne Water [7].

The project combines conventional drainage systems with WSUD measures at the streetscape and sub-catchment level, with the aim of attenuating and treating stormwater flows to protect receiving waters within the development. Primary treatment of the stormwater is carried out by grass swales and an underground gravel trench system, which collects, infiltrates, and conveys road/roof runoff. The main boulevard acts as a bioretention system with an underground gravel-filled trench to allow for infiltration and conveyance of stormwater. The catchment runoff then undergoes secondary treatment through a wetland system before discharge into an ornamental lake. This project is significant as the first residential WSUD development of this scale in Australia. Its performance in exceeding the Urban Storm water Best Practice Management Guidelines for Total Nitrogen, Total Phosphorus and Total Suspended Solids levels, has won it both the 2000 President's Award in the Urban Development Institute of Australia Awards for Excellence (recognising innovation in urban development), and the 2001 Cooperative Research Centres' Association Technology Transfer Award. Its success as a private-sector implemented WSUD system led to its proponent Urban and Regional Land Corporation (URLC) to look to incorporate WSUD as a standard practice across the State of Victoria. The project has also attracted attention from developers, councils, waterway management agencies and environmental policy-makers throughout the country [7].

2.2 Large-Scale Remediation for the Sydney 2000 Olympic Games

Homebush Bay, NSW

For the establishment of the Sydney 2000 Olympic Games site, the Brownfield area of Home Bush Bay was remediated from an area of landfill, abattoirs, and a navy armament depot into a multiuse Olympic site. A Water Reclamation and Management Scheme (WRAMS) was set up in 2000 for large-scale recycling of non-potable water [8], which included a range of WSUD technologies. These technologies were implemented with a particular focus on addressing the objectives of protecting receiving waters from storm water and wastewater discharges; minimising potable water demand; and protecting and enhancing habitat for threatened species 2006 [9]. The focus of WSUD technologies was directed toward the on-site treatment, storage, and recycling of storm water and wastewater. Storm water runoff is treated using gross pollutant traps, swales and/or wetland systems. This has contributed to a reduction of 90% in nutrient loads in the Haslams Creek wetland remediation area [8]. Wastewater is treated in a water reclamation plant. Almost 100% of sewage is treated and recycled [49]. The treated water from both storm water and wastewater sources is stored and recycled for use throughout the Olympic site in water features, irrigation, toilet flushing, and fire fighting capacities [9]. Through the use of WSUD technology, the WRAMS scheme has resulted in the conservation of 850 million litres (ML) of water annually [49], a potential 50% reduction in annual potable water consumption within the Olympic site [9], as well as the annual diversion of approximately 550 ML of sewage normally discharged through ocean outfalls [8]. As part of the long-term sustainability focus of the 'Sydney Olympic Park Master Plan 2030', the Sydney Olympic Park Authority (SOPA) has identified key best practice environmental sustainability approaches to include, the connection to recycled water and effective water demand management practices, maintenance and extension of recycled water systems to new streets as required, and maintenance and extension of the existing stormwater system that recycles water, promotes infiltration to subsoil, filters pollutants and sediments, and minimises loads on adjoining waterways [50]. The SOPA has used WSUD technology to ensure that the town remains 'nationally and internationally' recognised for excellence and innovation in urban design, building design and sustainability [50], both in the present and for future generations.

References

1. Minnamurra Avenue Water Sensitive Road Project (PDF). Archived from the original (PDF) on 13 Sep 2009. Retrieved 22 Sep 2011. Archived 29 Mar 2012, at the Wayback Machine
2. Cremorne street reconstruction constructed 2003 (PDF). Archived from the original (PDF) on 02 Apr 2012. Retrieved 22 Sep 2011
3. Fowler street reconstruction constructed 2003/04 (PDF). Archived from the original (PDF) on 02 Apr 2012. Retrieved 22 Sep 2011
4. Stawell street reconstruction constructed 2004 (PDF). Archived from the original (PDF) on 02 Apr 2012. Retrieved 22 Sep 2011
5. Melbourne Water (n.d.) Lynbrook Estate, viewed 20 Sep 2011
6. Sydney Olympic Park Authority (2011) Water & catchments. Sydney Olympic Park Authority, viewed 20 Sep 2011
7. Sydney Olympic Park Authority (2011) Sydney Olympic Park master plan 2030. Sydney Olympic Park Authority, viewed 20 Sep 2011
8. Department of Environment and Conservation NSW (2006) Managing urban stormwater: harvesting and reuse. Department of Environment and Conservation NSW 2006, Sydney, viewed 19 Sep 2011
9. Brown R, Clarke J (2007) The transition toward water sensitive urban design: the story of Melbourne, Australia, Report No. 07/01. Facility for Advancing Water Biofiltration, Monash University, Clayton, VIC, p 44

Countries That Recycle Wastewater into Drinking Water

Rishika Mukhopadhyay

It likes to call this "toilet-to-tap water". Word for word: water passing from your toilet bowl to your tap. It's a rather bland expression, which overlooks the many chemical, physical and biological processes involved in recycling wastewater into top-quality drinking water.

While this idea may still make many people wince, it is becoming a serious consideration for more and more countries and municipalities around the globe, which are rightly concerned about dwindling freshwater resources. The Indian megalopolis of Bangalore, for example, is working on recycling wastewater in this way, as is Los Angeles.

In the United States, the Texan cities of Big Springs and Wichita Falls are already using this technique, known as direct potable reuse, as has the South African town of Beaufort since 2011.

1 Namibia Has Been Treating Wastewater for Drinking Since 1968

But in this field, and this remains a little-known fact, it's yet another country that remains the undisputed pioneer: Namibia. At 139th place in the ranking of countries by level of development, it may come as a surprise that this southern African state is so much in the vanguard, but when you consider the near-absence of water resources in its capital, Windhoek, it's immediately less surprising that this municipality has sought to innovate.

R. Mukhopadhyay (✉)
Department of Civil Engineering, Elitte College of Engineering, MAKA University of Technology, Kolkata 700113, India
e-mail: mukherjee.rishika@gmail.com

B. G. Gupta (ed.), *Urban Development with Treated Recycled and Reuse of Wastewater: Sustainable Management*, Urban Sustainability,
https://doi.org/10.1007/978-981-96-0857-7_10

In the middle of an arid plain, more than 200 km from the coast, Windhoek (population 500,000) can neither harvest the scarce rainwater, which evaporates almost immediately in its desert latitudes, nor draw from the rivers or groundwater in the surrounding area, which are only recharged to a limited extent when they are not dried up.

In 1968, the city, then under South African domination, was also seeing its population grow at an impressive rate, when it began to recycle its wastewater and transform it into drinking water. Fifty-five years later, 30% of wastewater is recycled into drinking water in less than ten hours. The rest of the domestic drinking water comes from dams and boreholes in other parts of the country.

2 Potabilizing Wastewater in 10 Steps

To enable wastewater to be recycled as drinking water, Windhoek has set up an innovative sequence of processes that currently comprises **10 stages**. It includes physicochemical processes, such as coagulation and flocculation (addition of a coagulant to create flocs, i.e. groupings of suspended matter which then fall by their own weight and are eliminated in the sludge), as well as chemical processes such as ozonation.

In contact with ozone, the water undergoes an oxidation process that breaks down numerous micropollutants (pesticides, drug residues, etc.) and inactivates bacteria, viruses and parasites.

This is followed by the final stages of biological filtration on granular activated carbon and physical filtration (activated carbon filtration and membrane ultrafiltration) to eliminate any remaining soluble pollution. Before being sent into the water network, the water undergoes quality controls and chlorination, ensuring a disinfectant effect that lasts over time, so that the quality of the water obtained does not deteriorate during distribution.

In recent years, the Windhoek wastewater treatment plant has welcomed intrigued and interested **visitors** from **Australia**, Germany, the United Arab Emirates... and with good reason, the techniques developed in Namibia remain interesting in more ways than one.

3 A Less Costly Solution Than Desalination

For countries seeking new sources of drinking water, wastewater recycling is less energy-intensive and more environmentally friendly than seawater **desalination**, a technique that is more widespread worldwide. Where potabilization of wastewater consumes between 1 and 1.5 kWh per m^3, desalination requires between 3 and 4 kWh per m^3. What's more, desalination produces bulky waste products: salt and pollutant concentrates that are often discharged directly into the seas and oceans, disrupting ecosystems.

Conference organized by the Barcelona Metropolitan Area on the recycling of wastewater into drinking water in Namibia, with speakers Piet Du Pisani, in charge of wastewater management in Windhoek, and ASERA, a Spanish association promoting the reuse of wastewater.

Despite all these advantages, and the convincing results of the Namibian case study, the potabilization of wastewater is still in its infancy on a global scale, as its application involves overcoming a number of barriers. Firstly, the cost of installation. To date, only developed countries have been able to finance such projects, either on their own soil (in the United States, Singapore, etc.) or abroad, with, for example, the modernization of the Namibian plant, a **public–private partnership** involving the French company Véolia, the Australian-Indian company Wabag and the city of Berlin.

4 Wastewater Re-use

Water reuse if a political and socio-economic is a major issue for the future development of drinking water and wastewater services worldwide. Indeed, the major advantage it has is that an alternative and permanent resource is ensured allowing water shortages to be reduced, natural resources to be better conserved and water shortages caused by climate change to be alleviated. Some countries, states and large cities (Australia, California, Cyprus, Spain, Florida, Israel, Jordan, Malta, Singapore, etc.) already have ambitious objectives to meet 10 to 30%, even as much as 60%, of their water demand through reusing purified wastewater.

Agricultural irrigation was, is and will remain the largest consumer of recycled water with several well recognised advantages and benefits, in particular its contribution to food safety.

The reuse and internal recycling of industrial wastewater have become common practice for many industries with new trends such as the zero liquid discharge objective and the reuse of inter-sector water such as the use of urban wastewater for industrial purposes.

Water reuse in an urban environment is marked by a rapid development due to its crucial role in tomorrow's sustainable city. Moreover, the irrigation of urban green spaces and other applications are gaining importance and include industrial uses (cleaning, fire-fighting, cooling towers, etc.), recycling in buildings and environmental uses for maintaining and restoring water bodies, rivers and wetlands.

Water table recharge, feeding reservoirs for the indirect production of drinking water, even the direct reuse of ultra-pure water to increase water supply have been deployed as sustainable solutions against increasing water shortages in some countries who will face this problem with the next 20 years.

5 Irrigation in Agriculture and Green Spaces

Agriculture is the water's largest water consumer, i.e. around 70% of world demand. In some arid and semi-arid countries, the majority of irrigation water is recycled water. Over and above the usual factor of chronic water shortage, the need for an alternative water resource has accelerated over the last few years due to increasingly severe and repetitive droughts, which not only occur in traditionally arid areas in the United States, the Mediterranean region, the Middle East and South Asia, but also in a certain number of temperate regions, e.g. in Europe and in North America.

The reuse of wastewater in agriculture has been practised for thousands of years. It was developed by ancient civilisations and up until the twentieth century was also used as a system to purify wastewater in fields used for land spreading. Indeed soil represents an efficient filter with up to one or two tonnes of "purifying" microorganisms per hectare. Not only does wastewater provide water for crops but also contributes to improving the yield by providing a input in nutrients.

Currently, the main interest in reusing wastewater in agriculture is to alleviate in moisture and to increase agricultural production yields by a suitable supply of irrigation water. To maintain food production, technical and research work over the last few years has focused more on efficient irrigation by developing localised irrigation systems (drip irrigation) and/or computer-controlled networks equipped with sensors. However, efficient irrigation alone is not enough to resolve the problems of moisture deficits. Therefore, supply from an alternative resource through the reuse of wastewater is becoming a development priority in many countries and regions.

The underlying principle of water reuse for agricultural purposes is the need for the appropriate treatment of municipal wastewater to achieve a specific quality for a given use. It should be noted that besides the well-known benefits, using recycled water for irrigation purposes may have negative impacts on public health and the environment and which depend on the level of treatment, local conditions and irrigation practices. In all cases, existing scientific knowledge, operational feedback and best practices allow risks to be reduced through the implementation of efficient planning, the selection of appropriate technology and rigorous management of irrigation practices.

The main risks related to the reuse of treated wastewater for irrigation purposes may be divided into three categories:

- Health risks,
- Agronomic and environmental risks,
- Operational risks in the deterioration in the quality of recycled water within distribution systems and clogging of irrigation equipment.

On principle, most of the more recent standards require as a minimum, biological treatment of wastewater intended for reuse as irrigation water. In certain cases, such as Mexico for example, priority may be given to conserving the fertilising value of wastewater by an advanced primary treatment process (e.g. coagulation, flocculation and lamellar settling), followed by filtration and/or disinfection for priority removal of microbiological pollution, whilst keeping carbon and nutrients for irrigated crops.

Additional tertiary treatment is often essential for uses with high sanitary risks such as irrigated market produce eaten raw and the irrigation of green spaces. Additional filtration is also required in order to avoid deposits in the distribution system and/or the risk of clogging of irrigation equipment, especially of spray nozzles and drip irrigation systems.

The development and implementation of new wastewater treatment processes for irrigation has also contributed to improving the aesthetic quality of recycled water—through the removal of odour problems and the colour of recycled water—which were shown to be obstacles for several projects due to the negative perception by users.

6 Controlling Health Risks

The major public health risk associated with recycled water irrigation is the contamination by pathogenic micro-organisms, namely, viruses, bacteria, parasitic worms (helminths) and protozoa. As highlighted by the World Health Organisation (WHO) in its guidelines on the use of wastewater in agriculture dated 2006, a risk of transmitting infections exists when the following conditions are met:

- the population, the agricultural workers or animals are exposed to pathogens in recycled water by direct or indirect contact with them in irrigated areas or by consuming contaminated crops;
- the number of pathogens ingested is higher than the infective dose;
- the host becomes infected;
- the infection leads to an illness and /or transmission of the pathogen to other people or animals.

Consequently, the main objective of these public health protection measures in water projects, is to prevent the first two conditions from occurring. This means that the appropriate practices must be implemented in order to reduce the number of pathogens in recycled water as well as implementing several barriers and other measures to reduce the probability of coming into contact with potentially infectious micro-organisms. The choices in best practices for protecting human and animal health depends on local conditions and economic aspects.

The following four areas of best practice constitute the main measures to control health risks when irrigating with recycled water:

- wastewater treatment and quality control during water distribution and storage;
- audit on the use of wastewater through the choice of suitable irrigation methods and crop farming practices;
- restriction on the types of crops to be irrigated and restricted public access, in particular for irrigating green spaces;
- other measures, including controlling human exposure (e.g. measures to protect farm workers such as gloves, masks, etc.), harvesting methods, educating the public concerned and promoting best practices in health.

Treating wastewater is widely recognised as the most effective measure to reduce health risks linked to the use of recycled water for irrigation purposes. For this reason, all national or international regulations set or recommend the quality that wastewater treatment must achieve in accordance with the use envisaged and the level of risk involved. The degree of risk depends on several factors, including local endemic health, farming methods, weather conditions, the risk of direct contact, economic feasibility, etc.

Over the last few years several countries and federal states have included the recycling of water in their national water resources management policies with rules, obligations and financial incentives. In addition to the level of quality in relation to microbiological and physico-chemical indicators, certain standards also recommend additional treatment processes and barriers in order to reduce health risks.

In 2006, based on the approach used by the Australian Controlled Health Risk standards, the WHO reviewed all of its 1989 recommendations on agricultural irrigation using treated wastewater. The guide values were maintained at the same level, thereby allowing health risks to be limited to an acceptable level for the protection of public health, but with new concepts on additional barriers which ensure the economic viability of reuse projects.

The USA is considered as a world leader in wastewater reuse, with the first regulations adopted in California in 1918 followed by several reviews, the last dating from December 2000. At a federal level, the USEPA revised its 2004 recommendations in 2012 by mainly maintaining two recycling water quality guideline levels and recommendations on better monitoring of treatment processes.

In most of the standards and regulations, a precautionary principal is applied for high-risk uses, such as, for example, the irrigation of crops consumed raw and restriction-free irrigation of green spaces in the urban environment.

7 Controlling Agronomic and Environmental Risks

As a general rule, environmental risks are mainly agronomic chemical risks linked to the potential presence of trace elements, heavy metals and organic micropollutants in recycled water. Preventive measures for protection against health risks are more than sufficient to protect soil, surface water and ground water.

Environmental risks on the pollution of resource water are, in principle, taken into account in regulations protecting drinking water intake zones and at-risk areas.

Consequently, and as demonstrated by state-of-the-art studies, the major agronomic risks are as follows:

- excessive salinity of the soil which affects the transpiration and growth of sensitive cultures. The risk of soil becoming saline is high in arid zones and may be assessed by monitoring the concentration levels of dissolved solids, electrical conductivity and chlorides.
- excess sodium which deteriorates the structure of clay soil and which may cause its permeability to be reduced. In addition to an excessive quantity of exchangeable sodium, this harmful effect also encourages high **pH** and low electrical conductivity.
- toxic conditions for crops, in particular linked to high levels of boron, sodium and chlorides, and occasionally certain trace elements (including heavy metals), most often coming from industrial effluent. Toxicity affects plant growth, shown by signs of burning and/or leaf loss.
- excessive nitrogen can affect a crop's nutrient balance and the quality of surface or ground water.

Considering that wastewater treatment has no effect on chemical compounds, the best solution consists in controlling: saline effluent discharge at source, the intrusion of seawater in wastewater networks, the use of detergents containing boron and industrial discharges.

In order to control agronomic risks, best recycled water irrigation practices consist in combining several preventive or remedial measures, including:

- selecting an adequate irrigation method,

- choosing the right crops to irrigate, which are sufficiently tolerant to sodium and chlorides,
- suitable soil management practices (preparation, soil enhancers, etc.),
- sufficient leaching and drainage of soil in order to drain water and evacuate excess of salts,
- appropriate schedule to both irrigate and leach,
- correct use of fertilisers.

8 Controlling Recycled Water Quality in Distribution and Irrigation Systems

The main problem related to the distribution of purified wastewater is the potential deterioration of microbiological quality, in particular in hot climates and very long distribution networks. For this reason as well as for uses carrying high health risks, it is recommended that a residual chlorine level be maintained in order to prevent bacterial growth. However, the concentration level of residual chlorine must not exceed 0.5–1 mg/L in order to avoid the toxic effects on sensitive plants.

The presence of suspended solids in recycled water leads to risks of deposits building up in the networks, to deterioration in the water's taste and odour qualities and localised fouling of irrigation systems. In order to overcome these drawbacks, best practices in distribution network maintenance includes regular hydraulic purges, even the mechanical cleaning of significant deposits.

It should be noted that drip irrigation systems often require additional filtration following storage and specific cleaning protocols using chemicals.

9 Industrial Uses

One of the greatest potentials for water reuse is to add to or to replace the use of drinking water and/or natural resources in industry. Industry is the second biggest water consumer after agriculture and accounts for around 25% of world's demand.

In general, the reduction of water demand in industry, up to and including the end of industrial water cycle, includes three water-saving strategies and the minimisation of wastewater discharge:

- cascading reuse, involving direct reuse with little or not treatment;
- recycling of water following suitable treatment;
- wastewater reduction at source by reducing the water requirements of industrial processes.

Industrial wastewater reuse and internal recycling are well-established practices. Their development potential should increase in the future with the growth in water deficits and the supply costs of fresh water as well as increasingly stringent regulatory requirements regarding discharges.

Water reuse has been traditionally practised for years in the oil and gas, textile, car, paper and pulp and energy production industries and more recently in the electronic and food industries.

Although existing types of industrial water reuse are wide, the main uses are:

- Open or closed circuit cooling systems,
- Cleaning water,
- Boiler feed water,
- Process water,
- Various other uses such as fire-fighting, cleaning, etc.

The requirements and areas of application of industrial water recycling depend upon the industry, specific industrial processes as well as their efficiency targets. For this reason, it is impossible to generalise quality requirements for recycled water used as process water.

10 Cooling Water

When using recycled water in cooling towers, the major health concern is the risk of inhaling spray containing potentially pathogenic agents For this reason, the majority of regulations require that water be treated with an additional wastewater disinfection process.

Despite the very low concentration levels of micro-organisms in treated water, the risk of biological growth in cooling systems is very high due to the presence of nutrients and to high temperatures. In addition to health risks, bacterial development may produce unwelcome biofilms and deposits, which can alter the transfer of heat and cause microbiological corrosion and/or clogging of heat exchangers and spray nozzles. The method the most used to limit bacterial growth in cooling systems is the addition of biocides.

Another, potentially negative impact when using recycled water as cooling water is scaling. The scaling potential depends on concentration levels in calcium, magnesium, sulphate, phosphate, silica and fluoride as well as the level of alkalinity. The type of scaling most commonly found is caused by calcium phosphate, followed by silica and calcium sulphate (quite frequent).

11 Boiler Water

The use of recycled water for feeding boilers is not very different from the use of conventional drinking water because a demineralisation pre-treatment step is always necessary. For this usage, the quality of recycled water depends on the boiler's operating pressure, with those with the highest pressure requiring better quality water.

Accumulation of scale and the corrosion of equipment are the major risks for this type of use. To limit these risks, in addition to eliminating water hardness, alkalinity and dissolved solids, the concentration levels of calcium, magnesium and silica are also limited and controlled.

12 Diversification of Uses and Reuse of Urban Wastewater in Industry

Significant progress in wastewater treatment technologies, in particular the development of membrane processes, has encouraged industrial uses and accelerated work between sectors. Several urban wastewater reuse projects have been implemented over the last ten years, essentially for supplying cooling towers and boilers in the petrochemical industry and energy production sector. Since 2000 for example, several refineries in California have adopted the use of urban recycled water as their main source of boiler and cooling water.

The West Basin water recycling plant to the south of Los Angeles is one of the biggest reuse plants in the world (capacity of 220,000 m^3/day in 2015) and a world leader in the diversification in the uses of recycled water. For the first time ever, this plant deployed the "fit for purpose" water production and "designer water" concept, distributing five different water qualities, three of which are specifically intended for industries (1) nitrified cooling water, (2) demineralised water for low pressure boilers and (3) very highly quality demineralised water for high pressure boilers.

Recycled water production for supplying cooling towers is performed on-site by bio-filtering recycled water, a satellite treatment, which is distributed for irrigation purposes and other urban uses. Compact and high productivity Biofor® bioreactors are used for the total nitrification of water using residual ammonia concentration levels below the detection limit of 0.1 mgN/L (average input concentration level of 40 mgN/L).

Boiler feed water in refineries requires a much higher level of demineralisation (table 37), which is performed by a reverse osmosis treatment process following microfiltration pre-treatment. For example, the dissolved solids (TDS) in feed water are reduced from 800–900 ppm to around 50 ppm for low pressure boilers. For high pressure boiler feed water, a second passage through the reverse osmosis stage is performed in order to reduce dissolved solids to below 5 ppm.

13 Urban and Municipal Uses

Non-potable water distribution in the urban environment is quite an old activity (e.g. Los Angeles, Paris, San Francisco, etc.) in which particular interest has been shown over the last few years. In response to increasingly frequent droughts and water shortages, several cities (e.g. in Australia, China, Spain, United States, etc.) have developed a dual, recycled water distribution network for irrigation and other urban uses. The main water reuse categories in the urban environment are as follows:

- The irrigation of green spaces, which is the most common usage and which includes the irrigation of public and private parks, sports grounds, greenbelt land, golf courses as well as residential areas and private gardens.
- Other examples of urban uses include street cleaning, carwashes, fire-fighting, air-conditioning, toilet flush water and other commercial uses.
- Recycling within a building which is essentially water recycling in high-rise buildings, including offices, shopping malls and private residential blocks.
- Improving the environmental and recreational uses for the reconstitution and the supply of water bodies, including those used for swimming (with or without physical contact), leisure pursuits or fishing.

Generally, water reuse in the urban environment requires a suitable infrastructure and in particular a dual distribution network. Dual distribution and plumbing systems are relatively easy and economically viable to install in new urban areas and/or new buildings.

Protecting public health is the most important requirement for this type of use due to risks of direct contact with recycled water. Due to this, effluent disinfection requirements are the amongst the most stringent for non-drinking water uses and are comparable to those for restriction-free irrigation for parks and green spaces open to the public. In order to achieve such a treatment level with near-total disinfection, additional filtration and disinfection treatment processes are necessary following biological treatment. In addition to monitoring recycled water quality, stringent and regular control of the distribution systems is recommended, in particular in order to avoid all risks of interconnecting with the drinking water the network.

14 Potable Uses

Technological progress in the domain of water treatment enable excellent quality recycling water to be produced from urban wastewater, which may even be of better quality than the drinking water from springs. Several scientific studies have demonstrated that there are no pertinent objections for reusing purified wastewater as drinking water following appropriate treatment. However, the main constraints in this type of use are psychological and cultural, with a negative perception of wastewater as being unhealthy and irreversibly contaminated.

Although not publicly recognised, several cities around world are supplied by unplanned wastewater reuse schemes, with the water diluted a little or less by surface water for indirect drinking water production. Compared to this existing and widespread situation, planned reuse of wastewater to increase drinking water resources have a certain number of advantages which allow health risks to be reduced and the profitability of water and wastewater treatment to be improved.

The history of the success of reusing purified wastewater for producing drinking water started in the 1960s with an acquifer recharge project at Montebello Forebay, California (1962) by infiltration of recycled water and first direct reuse as drinking water at Windhoek in Namibia (1968). Since then, the use of these projects experiences and those of several other projects implemented since, has demonstrated the feasibility of this type of reuse and the absence of negative effects on public health.

The most common practices in indirect reuse of purified water for drinking water production includes:

- Indirect (via infiltration tanks) or direct (via injection well) replenishment of aquifers used for drinking water production or as a barrier against the intrusion of seawater or polluted water (e.g. river bank filtration),
- Replenishment of surface reservoirs used for drinking water production.

Due to the high health risks related to possible microbiological or chemical pollution, treatment processes are based on the ''multiple barrier" concept in order to achieve and to ensure a very high degree of effectiveness and reliability. Very stringent safety requirements for public health are not only met through adequate treatment but also by regularly and continually controlling water quality as well as each individual process' reliability and by deploying best project management practices.

14.1 Scarcity is the New Reality, But Proven Technologies Bring Results

Droughts are likely to be an increasing problem around the world. Climate change is both reducing precipitation in many regions and dumping too much water at once. Urbanization is increasing water demand beyond the capacity of existing systems and resources. What solutions are available to prepare for drought?

14.2 Storm Water Capture

To manage long dry stretches punctuated by intense rain events, there are several ways to capture precipitation during rainy periods and save it for more arid times.

Among those ways are traditional dams and reservoirs. Groundwater banking, managed aquifer recharge, and groundwater storage projects are also effective without many of the negatives associated with damming rivers. Watershed upgrades can even out river flows throughout the year. And there are urban planning measures like water-sensitive urban design and "sponge cities" that hold and percolate water into aquifers.

14.3 Wastewater Reuse

Water scarcity is a global issue, from the American West's long-term mega drought to the aridity in Taiwan that recently affected microchip supplies. It no longer makes sense to discard water after one use.

With water reuse, wastewater is commonly treated to nonpotable standards to create an abundant and cost-effective stream for landscape or agricultural irrigation, equipment or pavement washing, toilet flushing, industrial processes, and many more applications.

14.4 Desalination of Brackish Water and Seawater

Desalination turns salty water into fresh, drinkable water for municipal supplies, industry, and agriculture. While it's generally more expensive than wastewater reuse, advances in reverse osmosis technology are lowering prices. Simultaneously, demand is increasing the value of water, and the knowledge base is growing for reducing desalination's effects on ecosystems. Furthermore, modular equipment is making desalination more viable on smaller scales.

Desalination can be a lifeline for parched inland areas like in the American West. Some 800 times more brackish groundwater lies in United States aquifers than the amount of fresh groundwater the nation pumps every year, and it's much easier and cheaper to desalinate brackish water than seawater.

One Seven Seas Water Group project in Alice, Texas, for instance, will provide a new source of freshwater from the city's brackish aquifer at a lower cost than it is paying for water already. A build-own-operate-transfer arrangement as part of our Water-as-a-Service® offering is delivering that water with no initial capital outlay. The city will have a more resilient water source and break free of its distant, piped-in supply.

14.5 Complete Water Cycle Solutions

Complete water cycle solutions combine as many of these measures as possible. Given brackish or seawater, a desalination plant produces fresh drinking water. The wastewater is then converted to effluent that is safe for myriad nonpotable uses. Where storm water is captured, a water purification plant makes it safe to drink.

In a nutshell, a complete water cycle solution takes advantage of every drop of available water. Seven Seas has the knowledge and vast supplier network to implement a complete water cycle solution for maximum drought resilience. While facing down drought can be daunting, there are ways to power through and flourish even with no initial capital. Contact Seven Seas to learn more.

Sequence of reclamation from left: raw sewage, sewage treatment plant effluent, and finally reclaimed water (after several treatment steps).

Water Reclamation

Aftabuz Zaman

Water reclamation is the process of converting municipal wastewater or sewage and industrial wastewater into water that can be reused for a variety of purposes. It is also called **wastewater reuse**, **water reuse** or **water recycling**. There are many types of reuse. It is possible to reuse water in this way in cities or for irrigation in agriculture. Other types of reuse are environmental reuse, industrial reuse, and reuse for drinking water, whether planned or not. Reuse may include irrigation of gardens and agricultural fields or replenishing surface water and groundwater. This latter is also known as groundwater recharge. Reused water also serve various needs in residences such as toilet flushing, businesses, and industry. It is possible to treat wastewater to reach drinking water standards. Injecting reclaimed water into the water supply distribution system is known as direct potable reuse. Drinking reclaimed water is not typical [1]. Reusing treated municipal wastewater for irrigation is a long-established practice. This is especially so in arid countries. Reusing wastewater as part of sustainable water management allows water to remain an alternative water source for human activities. This can reduce scarcity. It also eases pressures on groundwater and other natural water bodies [2].

There are several technologies used to treat wastewater for reuse. A combination of these technologies can meet strict treatment standards and make sure that the processed water is hygienically safe, meaning free from pathogens. The following are some of the typical technologies: Ozonation, ultrafiltration, aerobic treatment (membrane bioreactor), forward osmosis, reverse osmosis, and advanced oxidation [3], or activated carbon [4]. Some water-demanding activities do not require high grade water. In this case, wastewater can be reused with little or no treatment.

The cost of reclaimed water exceeds that of potable water in many regions of the world, where fresh water is plentiful. The costs of water reclamation options might

A. Zaman (✉)
School of Agricultural Sciences, Sister Nivedita University, Kolkata, India
e-mail: profazaman@gmail.com

B. G. Gupta (ed.), *Urban Development with Treated Recycled and Reuse of Wastewater: Sustainable Management*, Urban Sustainability,
https://doi.org/10.1007/978-981-96-0857-7_11

be compared to the costs of alternative options which also achieve similar effects of freshwater savings, namely greywater reuse systems, rainwater harvesting and stormwater recovery, or seawater desalination.

Water recycling and reuse is of increasing importance, not only in arid regions but also in cities and contaminated environments [5]. Municipal wastewater reuse is particularly high in the Middle East and North Africa region, in countries such as the UAE, Qatar, Kuwait and Israel [6].

The term "water reuse" is generally used interchangeably with terms such as wastewater reuse, water reclamation, and water recycling. A definition by the USEPA states: "Water reuse is the method of recycling treated wastewater for beneficial purposes, such as agricultural and landscape irrigation, industrial processes, toilet flushing, and groundwater replenishing (EPA, 2004) [7, 8]." A similar description is: "Water Reuse, the use of reclaimed water from treated wastewater, has been a long-established reality in many (semi)arid countries and regions. It helps to alleviate water scarcity by supplementing limited freshwater resources [9]."

The water that is used as an input to the treatment and reuse processes can be from a variety of sources. Usually it is wastewater (domestic or municipal, industrial or agricultural wastewater) but it could also come from urban runoff.

A constructed wetland in Haran-Al-Awamied, Syria. Reclaimed water sign in Dunedin, Florida, United States. Reclaimed water is water that is used more than one time before it passes back into the natural water cycle. Advances in municipal wastewater treatment technology allow communities to reuse water for many different purposes. The water is treated differently depending upon the source and use of the water as well as how it gets delivered.

1 Driving Forces

The World Health Organization has recognized the following principal driving forces for municipal wastewater reuse [10, 11]:

1. increasing water scarcity and stress,
2. increasing populations and related food security issues,
3. increasing environmental pollution from improper wastewater disposal, and
4. increasing recognition of the resource value of wastewater, excreta and greywater.

In some areas, one driving force is also the implementation of advanced wastewater treatment for the removal of organic micropollutants, which leads to an overall improved water quality [4].

Water recycling and reuse is of increasing importance, not only in arid regions but also in cities and contaminated environments [5].

Already, the groundwater aquifers that are used by over half of the world population are being over-drafted [12]. Reuse will continue to increase as the world's population becomes increasingly urbanized and concentrated near coastlines, where local freshwater supplies are limited or are available only with large capital expenditure

[13, 14]. Large quantities of freshwater can be saved by municipal wastewater reuse and recycling, reducing environmental pollution and improving carbon footprint [5]. Reuse can be an alternative water supply option.

Achieving more sustainable sanitation and wastewater management will require emphasis on actions linked to resource management, such as wastewater reuse or excreta reuse that will keep valuable resources available for productive uses [2]. This in turn supports human wellbeing and broader sustainability.

2 Potential Benefits

Water/wastewater reuse, as an alternative water source, can provide significant economic, social and environmental benefits, which are key motivators for implementing such reuse programs. These benefits include [14, 15]:

- For Cities and Households: Increased water availability (drinking water substitution—keep drinking water for drinking and reclaimed water for non-drinking use such as industry, cleaning, irrigation, domestic uses, and toilet flushing).
- For the Environment: Reduced nutrient loads to receiving waters (i.e. rivers, canals and other surface water resources); reduced over-abstraction of surface and groundwater; enhanced environmental protection by restoration of streams, wetlands and ponds; reduced energy consumption associated with production, treatment, and distribution of water (1.2 to 2.1 kWh/m^3) [16] compared to using deep groundwater resources, water importation or desalination
- Reduced manufacturing costs of using high quality reclaimed water
- In agriculture: Irrigation with treated wastewater may contribute to improve production yields, reduce the ecological footprint and promote socioeconomic benefits [17, 18]. It may also lead to reduced application of fertilizers (i.e. conservation of nutrients and reducing the need for artificial fertilizer through soil nutrition by the nutrients existing in the treated effluents) [17].

Reclaiming water for reuse applications instead of using freshwater supplies can be a water-saving measure. When used water is eventually discharged back into natural water sources, it can still have benefits to ecosystems, improving streamflow, nourishing plant life and recharging aquifers, as part of the natural water cycle [19].

3 Scale

Global treated wastewater reuse is estimated at 40.7 billion m3 per year, representing approximately 11% of the total domestic and manufacturing wastewater produced [6]. Municipal wastewater reuse is particularly high in the Middle East and North Africa region, in countries such as the UAE, Qatar, Kuwait and Israel [6].

For the Sustainable Development Goal 6 by the United Nations, Target 6.3 states "Halving the proportion of untreated wastewater and substantially increasing recycling and safe reuse globally by 2030" [20].

4 Types and Applications

Treated wastewater can be reused in industry (for example in cooling towers), in artificial recharge of aquifers, in agriculture, and in the rehabilitation of natural ecosystems (for example in wetlands). The main reclaimed water applications in the world are shown below [21–23]:

Categories of use	Uses
Urban uses	Irrigation of public parks, sporting facilities, private gardens, roadsides; Street cleaning; Fire protection systems; Vehicle washing; Toilet flushing; Air conditioners; Dust control
Agricultural uses	Food crops not commercially processed; Food crops commercially processed; Pasture for milking animals; Fodder; Fibre; Seed crops; Ornamental flowers; Orchards; Hydroponic culture; Aquaculture; Greenhouses; Viticulture
Industrial uses	Processing water; Cooling water; Recirculating cooling towers; Washdown water; Washing aggregate; Making concrete; Soil compaction; Dust control
Recreational uses	Golf course irrigation; Recreational impoundments with/without public access (e.g. fishing, boating, bathing); Aesthetic impoundments without public access; Snowmaking
Environmental uses	Aquifer recharge; Wetlands; Marshes; Stream augmentation; Wildlife habitat; Silviculture
Potable uses	Aquifer recharge for drinking water use; Augmentation of surface drinking water supplies; Treatment until drinking water quality

5 Urban Reuse

In rarer cases reclaimed water is also used to augment drinking water supplies. Most of the uses of water reclamation are non-potable uses such as washing cars, flushing toilets, cooling water for power plants, concrete mixing, artificial lakes, irrigation for golf courses and public parks, and for hydraulic fracturing. Where applicable, systems run a dual piping system to keep the recycled water separate from the potable water.

Usage types are distinguished as follows:

- Unrestricted: The use of reclaimed water for non-potable applications in municipal settings, where public access is not restricted.

- Restricted: The use of reclaimed water for non-potable applications in municipal settings, where public access is controlled or restricted by physical or institutional barriers, such as fencing, advisory signage, or temporal access restriction [14].

6 Agricultural Reuse

Irrigation with recycled municipal wastewater can also serve to fertilize plants if it contains nutrients, such as nitrogen, phosphorus and potassium. There are benefits of using recycled water for irrigation, including the lower cost compared to some other sources and consistency of supply regardless of season, climatic conditions and associated water restrictions. When reclaimed water is used for irrigation in agriculture, the nutrient (nitrogen and phosphorus) content of the treated wastewater has the benefit of acting as a fertilizer [24]. This can make the reuse of excreta contained in sewage attractive [10].

The irrigation water can be used in different ways on different crops, such as for food crops to be eaten raw or for crops which are intended for human consumption to be eaten raw or unprocessed. For processed food crops: crops which are intended for human consumption not to be eaten raw but after food processing (i.e. cooked, industrially processed) [25]. It can also be used on crops which are not intended for human consumption (e.g. pastures, forage, fiber, ornamental, seed, forest and turf crops) [26].

7 Risks in Agricultural Reuse

In developing countries, agriculture is increasingly using untreated municipal wastewater for irrigation—often in an unsafe manner. Cities provide lucrative markets for fresh produce, so they are attractive to farmers. However, because agriculture has to compete for increasingly scarce water resources with industry and municipal users, there is often no alternative for farmers but to use water polluted with urban waste directly to water their crops.

There can be significant health hazards related to using untreated wastewater in agriculture. Municipal wastewater can contain a mixture of chemical and biological pollutants. In low-income countries, there are often high levels of pathogens from excreta. In emerging nations, where industrial development is outpacing environmental regulation, there are increasing risks from inorganic and organic chemicals. The World Health Organization developed guidelines for safe use of wastewater in 2006 [10], advocating a 'multiple-barrier' approach wastewater use, for example by encouraging farmers to adopt various risk-reducing behaviors. These include ceasing irrigation a few days before harvesting to allow pathogens to die off in the sunlight; applying water carefully so it does not contaminate leaves likely to be eaten raw;

cleaning vegetables with disinfectant; or allowing fecal sludge used in farming to dry before being used as a human manure [24].

Drawbacks or risks often mentioned include the content of potentially harmful substances such as bacteria, heavy metals, or organic pollutants (including pharmaceuticals, personal care products and pesticides). Irrigation with wastewater can have both positive and negative effects on soil and plants, depending on the composition of the wastewater and on the soil or plant characteristics [27].

8 Environmental Reuse

The use of reclaimed water to create, enhance, sustain, or augment water bodies including wetlands, aquatic habitats, or stream flow is called "environmental reuse" [14]. For example, constructed wetlands fed by wastewater provide both wastewater treatment and habitats for flora and fauna [citation needed].

9 Industrial Reuse

Treated wastewater can be reused in industry (for example in cooling towers).

10 Planned Potable Reuse

Planned potable reuse is publicly acknowledged as an intentional project to recycle water for drinking water. There are two ways in which potable water can be delivered for reuse—"Indirect Potable Reuse" (IPR) and "Direct Potable Reuse". Both these forms of reuse are described below, and commonly involve a more formal public process and public consultation program than is the case with de facto or unacknowledged reuse [14, 28].

Some water agencies reuse highly treated effluent from municipal wastewater or resource recovery plants as a reliable, drought-proof source of drinking water. By using advanced purification processes, they produce water that meets all applicable drinking water standards. System reliability and frequent monitoring and testing are imperative to their meeting stringent controls [3].

The water needs of a community, water sources, public health regulations, costs, and the types of water infrastructure in place—such as distribution systems, man-made reservoirs, or natural groundwater basins—determine if and how reclaimed water can be part of the drinking water supply. Some communities reuse water to replenish groundwater basins. Others put it into surface water reservoirs. In these instances the reclaimed water is blended with other water supplies and/or sits in storage for a certain amount of time before it is drawn out and gets treated again at a water treatment or distribution system. In some communities, the reused water is put directly into pipelines that go to a water treatment plant or distribution system.

Modern technologies such as reverse osmosis and ultraviolet disinfection are commonly used when reclaimed water will be mixed with the drinking water supply [3].

Many people associate a feeling of disgust with reclaimed water and 13% of a survey group said they would not even sip it [29]. Nonetheless, the main health risk for potable use of reclaimed water is the potential for pharmaceutical and other household chemicals or their derivatives (environmental persistent pharmaceutical pollutants) to persist in this water [30]. This would be less of a concern if human excreta was kept out of sewage by using dry toilets or, alternatively, systems that treat blackwater separately from greywater.

11 Indirect Potable Reuse

Indirect potable reuse (IPR) means the water is delivered to the consumer indirectly. After it is purified, the reused water blends with other supplies and/or sits a while in some sort of storage, man-made or natural, before it gets delivered to a pipeline that leads to a water treatment plant or distribution system. That storage could be a groundwater basin or a surface water reservoir.

Some municipalities are using and others are investigating IPR of reclaimed water. For example, reclaimed water may be pumped into (subsurface recharge) or percolated down to (surface recharge) groundwater aquifers, pumped out, treated again, and finally used as drinking water. This technique may also be referred to as *groundwater recharging*. This includes slow processes of further multiple purification steps via the layers of earth/sand (absorption) and microflora in the soil (biodegradation).

IPR or even unplanned potable use of reclaimed wastewater is used in many countries, where the latter is discharged into groundwater to hold back saline intrusion in coastal aquifers. IPR has generally included some type of environmental buffer, but conditions in certain areas have created an urgent need for more direct alternatives [31].

IPR occurs through the augmentation of drinking water supplies with municipal wastewater treated to a level suitable for IPR followed by an environmental buffer (e.g. rivers, dams, aquifers, etc.) that precedes drinking water treatment. In this case, municipal wastewater passes through a series of treatment steps that encompasses membrane filtration and separation processes (e.g. MF, UF and RO), followed by an advanced chemical oxidation process (e.g. UV, UV + H_2O_2, ozone) [14]. In 'indirect' potable reuse applications, the reclaimed wastewater is used directly or mixed with other sources.

12 Direct Potable Reuse

Direct potable reuse (DPR) means the reused water is put directly into pipelines that go to a water treatment plant or distribution system. Direct potable reuse may occur with or without ''engineered storage" such as underground or above ground tanks [14]. In other words, DPR is the introduction of reclaimed water derived from domestic wastewater after extensive treatment and monitoring to assure that strict water quality requirements are met at all times, directly into a municipal water supply system.

13 Reuse in Space Stations

Wastewater reclamation can be especially important in relation to human spaceflight. In 1998, NASA announced it had built a human waste reclamation bioreactor designed for use in the International Space Station and a crewed Mars mission. Human urine and feces are input into one end of the reactor and pure oxygen, pure water, and compost (humanure) are output from the other end. The soil could be used for growing vegetables, and the bioreactor also produces electricity [32, 33].

Aboard the International Space Station, astronauts have been able to drink recycled urine due to the introduction of the ECLSS system. The system costs $250 million and has been working since May 2009. The system recycles wastewater and urine back into potable water used for drinking, food preparation, and oxygen generation. This cuts back on the need to frequently resupply the space station [34].

14 De Facto Wastewater Reuse (Unplanned Potable Reuse)

De facto, unacknowledged or unplanned potable reuse refers to situations where reuse of treated wastewater is practiced but is not officially recognized [14]. For example, a sewage treatment plant from one city may be discharging effluents to a river which is used as a drinking water supply for another city downstream.

Unplanned Indirect Potable Use [35] has existed for a long time. Large towns on the River Thames upstream of London (Oxford, Reading, Swindon, Bracknell) discharge their treated sewage ("non-potable water") into the Thames, which supplies water to London downstream. In the United States, the Mississippi River serves as both the destination of sewage treatment plant effluent and the source of potable water.

15 Design Considerations

15.1 Distribution

A lavender-colored pipeline carrying non-potable water in a dual piping system in Mountain View, California, US

Non-potable reclaimed water is often distributed with a dual piping network that keeps reclaimed water pipes completely separate from potable water pipes.

16 Treatment Processes

There are several technologies used to treat wastewater for reuse. A combination of these technologies can meet strict treatment standards and make sure that the processed water is hygienically safe, meaning free from pathogens. Some common technologies include ozonation, ultrafiltration, aerobic treatment (membrane bioreactor), forward osmosis, reverse osmosis, advanced oxidation [3] or activated carbon [4]. Reclaimed water providers use multi-barrier treatment processes and constant monitoring to ensure that reclaimed water is safe and treated properly for the intended end use.

Some water-demanding activities do not require high grade water. In this case, wastewater can be reused with little or no treatment. One example of this scenario is in the domestic environment where toilets can be flushed using greywater from baths and showers with little or no treatment.

In the case of municipal wastewater, the wastewater must pass through numerous sewage treatment process steps before it can be used. Steps might include screening, primary settling, biological treatment, tertiary treatment (for example reverse osmosis), and disinfection.

Wastewater is generally treated to only secondary level treatment when used for irrigation.

A pump station distributes reclaimed water to users around a city. These may include golf courses, agricultural uses, cooling towers, or landfills.

16.1 Alternative Options

Rather than treating municipal wastewater for reuse purposes, other options can achieve similar effects of freshwater savings:

- Greywater Reuse Systems—At a household level, treated or untreated greywater may be used for flush toilets or to water a garden.
- Rainwater Harvesting and Stormwater Recovery—Urban design systems which incorporate rainwater harvesting and reduce runoff are known as water-sensitive urban design (WSUD) in Australia, low-impact development (LID) in the United States and sustainable urban drainage systems (SUDS) in the United Kingdom.
- Seawater Desalination—An energy-intensive process where salt and other minerals are removed from seawater to produce potable water for drinking and irrigation, typically through membrane filtration (reverse osmosis) or steam distillation.

16.2 Costs

The cost of reclaimed water exceeds that of potable water in many regions of the world, where fresh water is plentiful. However, reclaimed water is usually sold to citizens at a cheaper rate to encourage its use. As fresh water supplies become limited from distribution costs, increased population demands, or climate change, the cost ratios will evolve also. The evaluation of reclaimed water needs to consider the entire water supply system, as it may bring important flexibility into the overall system [36].

Reclaimed water systems usually require a dual piping network, often with additional storage tanks, which adds to the costs of the system.

16.3 Barriers to Implementation

Barriers to water reclamation may include:

- Full-scale implementation and operation of water reuse schemes still face regulatory, economic, social and institutional challenges [37].
- Low economic viability of water reuse schemes [37]. This may partly be due to costs of water quality monitoring and identification of contaminants [38]. Difficulties in contaminant identification may include the separation of inorganic and organic pollutants, microorganisms, colloids, and others [39]. Full cost recovery from water reuse schemes is difficult. There is a lack of financial water pricing systems comparable to already subsidized conventional treatment plants [40].
- Psychological barriers, sometimes referred to as the "yuck factor", can also be an impediment to implementation, particularly for direct potable reuse plans. These psychological factors are closely associated with disgust, specifically pathogen avoidance [41].

17 Health Aspects

Reclaimed water is considered safe when appropriately used. Reclaimed water planned for use in recharging aquifers or augmenting surface water receives adequate and reliable treatment before mixing with naturally occurring water and undergoing natural restoration processes. Some of this water eventually becomes part of drinking water supplies.

A study published in 2009 compared the differences in water quality between reclaimed/recycled water, surface water, and groundwater [42]. Results indicated that reclaimed water, surface water, and groundwater are more similar than dissimilar with regard to constituents. The researchers tested for 244 representative constituents typically found in water. When detected, most constituents were in the parts-per-billion and parts-per-trillion range. DEET (an insect repellant) and caffeine were found in all water types and in virtually all samples. Triclosan (in antibacterial soap and toothpaste) was found in all water types, but detected in higher levels (parts-per-trillion) in reclaimed water than in surface or groundwater. Very few hormones/steroids were detected in samples, and when detected were at very low levels. Haloacetic acids (a disinfection by-product) were found in all types of samples, even groundwater. The largest difference between reclaimed water and the other waters appears to be that reclaimed water has been disinfected and thus has disinfection byproducts (due to chlorine use).

A 2005 study found that there had been no instances of illness or disease from either microbial pathogens or chemicals, and the risks of using reclaimed water for irrigation are not measurably different from irrigation using potable water [43].

A 2012 study conducted by the National Research Council in the United States found that the risk of exposure to certain microbial and chemical contaminants from drinking reclaimed water does not appear to be higher than the risk experienced in some current drinking water treatment systems, and may be orders of magnitude lower [44]. This report recommends adjustments to the federal regulatory framework that could enhance public health protection for both planned and unplanned (or de facto reuse) and increase public confidence in water reuse.

17.1 Environmental Aspects

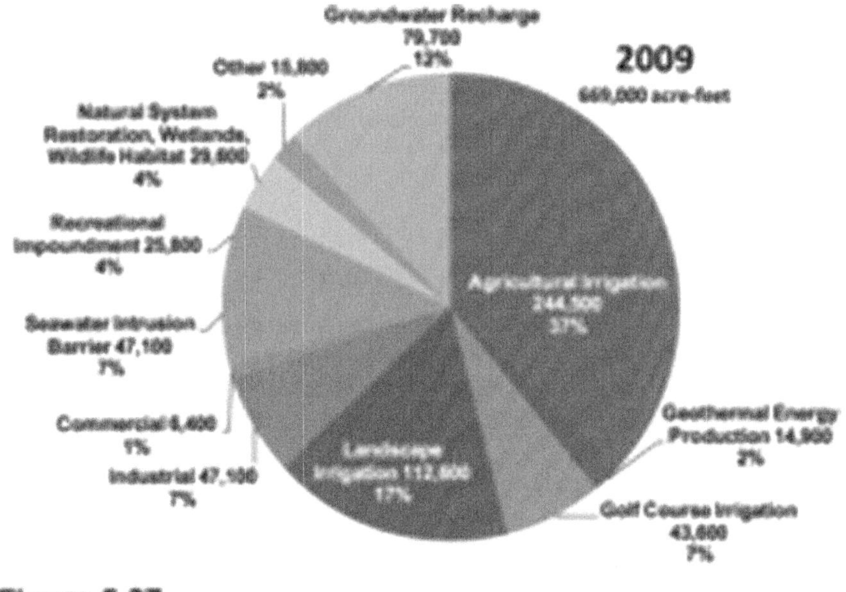

Figure 5-37
Uses of recycled water in Calif. (SWRCB 2011)

Using reclaimed water for non-potable uses saves potable water for drinking, since less potable water will be used for non-potable uses [45]

It sometimes contains higher levels of nutrients such as nitrogen, phosphorus and oxygen which may help fertilize garden and agricultural plants when used for irrigation [citation needed].

Fresh water makes up less than 3% of the world's water resources, and just 1% of that is readily available. Even though fresh water is scarce, just 3% of it is extracted for human consumption. The remaining water is mostly used for agriculture, which uses roughly two-thirds of all fresh water [46–48].

Reclaimed water can offer a viable and effective alternative to freshwater where freshwater supplies are scarce. Reclaimed water is utilized to maintain or increase lake levels, restore wetlands, and restore river flows during hot weather and droughts, protecting biodiversity. Additionally, reclaimed water is utilized for street cleaning, irrigation of urban green spaces, and industrial processes. Reclaimed water has the advantage of being a consistent source of water supply that is unaffected by seasonal droughts and weather changes [47–49].

The usage of water reclamation decreases the pollution sent to sensitive environments. It can also enhance wetlands, which benefits the wildlife depending on that ecosystem. It also helps to reduce the likelihood of drought as recycling of water reduces the use of fresh water supply from underground sources. For instance, the San Jose/Santa Clara Water Pollution Control Plant instituted a water recycling program to protect the San Francisco Bay area's natural salt water marshes [45].

The main potential risks that are associated with reclaimed wastewater reuse for irrigation purposes when the treatment is not adequate are the following [15, 21]:

1. Contamination of the food chain with microcontaminants, pathogens (i.e. bacteria, viruses, protozoa, helminths), or antibiotic resistance determinants;
2. Soil salinization and accumulation of various unknown constituents that might adversely affect agricultural production;
3. Distribution of the indigenous soil microbial communities;
4. Alteration of the physicochemical and microbiological properties of the soil and contribution to the accumulation of chemical/biological contaminants (e.g. heavy metals, chemicals (i.e. boron, nitrogen, phosphorus, chloride, sodium, pesticides/herbicides), natural chemicals (i.e. hormones), contaminants of emerging concern (CECs) (i.e. pharmaceuticals and their metabolites, personal care products, household chemicals and food additives and their transformation products), etc.) in it and subsequent uptake by plants and crops;
5. Excessive growth of algae and vegetation in canals carrying wastewater (i.e. eutrophication);
6. Groundwater quality degradation by the various reclaimed water contaminants, migrating and accumulating in the soil and aquifers.

18 Guidelines and Regulations

18.1 International Organizations

- World Health Organization (WHO): "Guidelines for the safe use of wastewater, excreta and greywater" (2006) [10].
- United Nations Environment Programme (UNEP): "Guidelines for municipal wastewater reuse in the Mediterranean region" (2005).
- United Nations Water Decade Programme on Capacity Development (UNW-DPC): Proceedings on the UNWater project "Safe use of wastewater in agriculture" (2013).

18.2 European Union

Since 26 June 2023 [50] there is an EU regulation on minimum requirements for water reuse for irrigation purposes [51]. The water quality requirements are divided into four categories depending on what is irrigated and how the irrigation is performed. The water quality parameters included are *E.coli*, BOD5, total suspended solids (TSS), turbidity, legionella, and intestinal nematodes (helminth eggs).

In the Water Framework Directive, reuse of water is mentioned as one of the possible measures to achieve the Directive's quality goals. However, this remains a relatively vague recommendation rather than a requirement: Part B of Annex VI refers to reuse as one of the "supplementary measures which Member States within each river basin district may choose to adopt as part of the programme of measures required under Article 11(4)" [52].

Besides that, Article 12 of the Urban Wastewater Treatment Directive concerning the reuse of treated wastewater states that "treated wastewater shall be reused whenever appropriate", which some consider not specific enough to promote water reuse as it may leave too much room for interpretation as to what can be considered as an "appropriate" situation to reuse treated wastewater.

Despite the lack of common water reuse criteria at the EU level, several member states have issued their own legislative frameworks, regulations, or guidelines for different water reuse applications (e.g. Cyprus, France, Greece, Italy, and Spain).

However, an evaluation carried out by the European Commission on the water reuse standards of several member states concluded that they differed in their approach. There are important differences among the standards regarding permitted uses, parameters to be monitored, and limit values allowed. This lack of harmonization among water reuse standards could potentially create trade barriers for agricultural goods irrigated with reclaimed water. Once on the common market, the level of safety in the producing member states may be not considered sufficient by the importing countries [38]. The most representative standards on wastewater reuse from European member states are the following [52]:

- Cyprus: Law 106 (I) 2002 Water and Soil pollution control and associated regulations (KDP 772/2003, KDP 269/2005) (Issuing Institutions: Ministry of Agriculture, Natural resources and Environment, Water Development Department).
- France: Jorf num.0153, 4 July 2014. Order of 2014, related to the use of water from treated urban wastewater for irrigation of crops and green areas (Issuing Institutions: Ministry of Public Health, Ministry of Agriculture, Food and Fisheries, Ministry of Ecology, Energy and Sustainability).
- Greece: CMD No 145116. Measures, limits and procedures for reuse of treated wastewater (Issuing Institutions: Ministry of Environment, Energy and Climate Change).
- Italy: DM 185/2003. Technical measures for reuse of wastewater (Issuing Institutions: Ministry of Environment, Ministry of Agriculture, Ministry of Public Health).

- Portugal: NP 4434 2005. Reuse of reclaimed urban water for irrigation (Issuing Institutions: Portuguese Institute for Quality).
- Spain: RD 1620/2007. The legal framework for the reuse of treated wastewater (Issuing Institutions: Ministry of Environment, Ministry of Agriculture, Food and Fisheries, Ministry of Health).

By 2023, a new EU agriculture law may raise water reuse by six times, from 1.7 billion m^3 to 6.6 billion m^3, and cut water stress by 5% [46, 53].

18.3 United States

In the U.S., the Clean Water Act of 1972 mandated elimination of the discharge of untreated waste from municipal and industrial sources to make water safe for fishing and recreation. The US federal government provided billions of dollars in grants for building sewage treatment plants around the country. Modern treatment plants, usually using oxidation and/or chlorination in addition to primary and secondary treatment, were required to meet certain standards [54] [clarification needed].

Los Angeles County's sanitation districts started providing treated wastewater for landscape irrigation in parks and golf courses in 1929. The first reclaimed water facility in California was built at San Francisco's Golden Gate Park in 1932. The Water Replenishment District of Southern California was the first groundwater agency to obtain permitted use of recycled water for groundwater recharge in 1962.

Denver's Direct Potable Water Reuse Demonstration Project [55] examined the technical, scientific, and public acceptance aspects of DPR from 1979 to 1993. A chronic lifetime whole-animal health effects study on the 1 MGD advanced treatment plant product was conducted in conjunction with a comprehensive assessment of the chemical and microbiological water quality. The $30 million study found that the water produced met all health standards and compared favorably with Denver's high quality drinking water. Further, the projected cost was lower than estimates for obtaining distant new water supplies.

Reclaimed water is not regulated by the U.S. Environmental Protection Agency (EPA), but the EPA has developed water reuse guidelines that were most recently updated in 2012 [56, 57]. The EPA Guidelines for Water Reuse represents the international standard for best practices in water reuse. The document was developed under a Cooperative Research and Development Agreement between the EPA, the U.S. Agency for International Development (USAID), and the global consultancy CDM Smith. The Guidelines provide a framework for states to develop regulations that incorporate the best practices and address local requirements.

This section is an excerpt from Water supply and sanitation in the United States § Water reuse].

Reuse of reclaimed water is an increasingly common response to water scarcity in many parts of the United States. Reclaimed water is being reused directly for various non-potable uses in the United States, including urban landscape irrigation of parks, school yards, highway medians and golf courses; fire protection; commercial uses such as vehicle washing; industrial reuse such as cooling water, boiler water and process water; environmental and recreational uses such as the creation or restoration of wetlands; as well as agricultural irrigation [58]. In some cases, such as in Irvine Ranch Water District in Orange County, it is also used for flushing toilets [59].

It was estimated that in 2002 a total of 1.7 billion US gallons (6,400,000 m^3) per day, or almost 3% of public water supply, were being directly reused. California reused 0.6 and Florida 0.5 billion US gallons (1,900,000 m^3) per day respectively. Twenty-five states had regulations regarding the use of reclaimed water in 2002 [58]. Planned direct reuse of reclaimed water was initiated in 1932 with the construction of a reclaimed water facility at San Francisco's Golden Gate Park. Reclaimed water is typically distributed with a color-coded dual piping network that keeps reclaimed water pipes completely separate from potable water pipes [60].

18.4 Trade Associations

- The **Water Reuse Association** is a trade association in the United States which promotes reuse of water. According to their website, "The Water Reuse Association is the nation's only trade association solely dedicated to advancing laws, policy, funding, and public acceptance of recycled water. Water Reuse represents a coalition of utilities that recycle water, businesses that support the development of recycled water projects, and consumers of recycled water [61]." The Water Reuse Research Foundation was merged into the WateReuse Association on July 11, 2016 [62].

18.5 Other Countries

- Canada: "Canadian guidelines for domestic reclaimed water for use in toilet and urinal flushing" (2010).
- China: China National Reclaimed Water Quality Standard; China National Standard GB/T 18920–2002, GB/T 19923–2005, GB/T 18921–2002, GB 20922–2007 and GB/T 19772–2005.
- Israel: Ministry of Health regulation (2005).
- Japan: National Institute for Land and Infrastructure Management: Report of the Microbial Water Quality Project on Treated Sewage and Reclaimed Wastewater (2008).
- Jordan: Jordanian technical base n. 893/2006 Jordan water reuse management Plan (policy).
- Mexico: Mexican Standard NOM-001-ECOL-1996 governing wastewater reuse in Agriculture.

- South Africa: The latest revision of the Water Services Act of 1997 relating to grey-water and treated effluent (Department of Water Affairs and Forestry, 2001).
- Tunisia: Standard for the use of treated wastewater in agriculture (NT 106–109 of 1989) and list of crops that can be irrigated with treated wastewater (Ministry of Agriculture, 1994).
- Australia: National level Guidelines: Government of Australia (the Natural Resource Management Ministerial Council, the Environment Protection and Heritage Council, and the Australian Health Ministers Conference (NRMMC-EPHC-AHMC)): "Guidelines for water recycling: managing health and environmental risks" Phase 1, 2006.[52]

Planned or unplanned, wastewater reuse is a long-standing human habit that has deep roots in the evolution of sanitation [63].

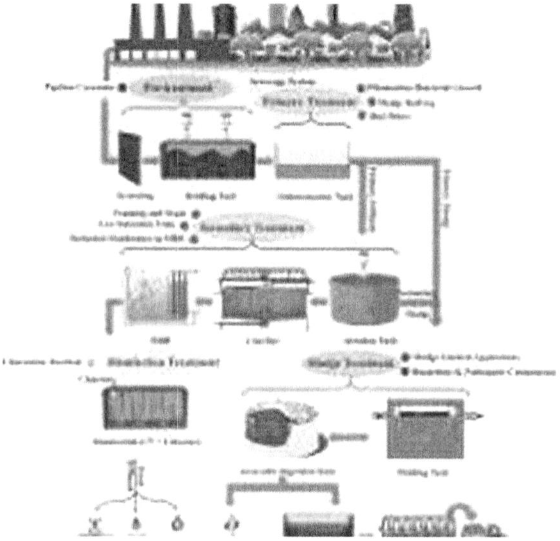

Water reuse system

19 Country Examples

19.1 Australia

This section is an excerpt from Water supply and sanitation in Australia § Wastewater reuse.

Interest in recovered wastewater possibilities rises during Australian droughts. Adelaide and Brisbane, two of Australia's largest cities, have both promised to augment their diminishing dams using recovered wastewater. In preparation for any future water scarcity, the former has also constructed a desalination facility. Upon completion, other cities and towns will examine the Western Corridor Recycled Water Project, which has established Brisbane as a leader in this trend [64, 65]. Regional Victoria, Australia, Goulbourn, Canberra, and Newcastle are all thinking about constructing a recovered effluent process. Communities in remote areas of Goulburn, New South Wales (NSW), the Australian Capital Territory (ACT), and Toowoomba, Queensland have been contemplating indirect potable reuse (IPR).

19.2 Israel

This section is an excerpt from water supply and sanitation in Israel § reclaimed water.

As of 2010, Israel leads the world in the proportion of water it recycles [66]. In Israel, 80% of the sewage, which amounts to 400 billion litres per year, is treated and utilised as irrigation water for public works and agriculture. In the Tel Aviv metropolitan region, 100% of the sewage is treated and reused. The UN recognised the Dan Region Wastewater Treatment Plant in 2012 as an example for the world. The facility, which goes by the name Shafdan in the area, gained praise for its innovative approach to sewage treatment, which was based on the natural filtering properties of sand [67, 68]. The agricultural sector accounted for the vast majority of the 400 million cubic metres of treated wastewater reused in 2010 [69]. Over 40% of the water used in farming comes from this source.

19.3 Namibia

This section is an excerpt from Water supply and sanitation in Namibia § Water reuse.

Many metropolitan regions in Namibia, including Swakopmund, Walvis Bay, Tsumeb, Otjiwarongo, Okahandja, Mariental, Oranjemund, and Windhoek, use the reuse of treated wastewater. Irrigation water reuse is common in most areas. Windhoek also makes use of recovered water for drinking purposes. Over the last half-century, Windhoek has made use of recycled water. Reclaimed water makes up almost 30% of the city's current drinking water supply, which serves 400,000 people [46, 70, 71]. Windhoek (Namibia, New Goreangab Water Reclamation Plant, NGWRP) is a prime example of direct potable reuse because it has been mixing treated wastewater with drinking water for over 45 years. This process is based on the multiple treatment barriers concept, which includes pre-ozonation, enhanced coagulation/dissolved air flotation/rapid sand filtration, and subsequent ozone, biological activated carbon/granular activated carbon, ultrafiltration (UF), chlorination, in order to decrease risks and improve water quality [72, 73]. Since the year 1968 the capital of

Namibia, Windhoek, has used reclaimed wastewater as one of their drinking water sources [74], which nowadays represent about 14% of the city's drinking water production [75]. The City of Windhoek constructed the New Goreangab Reclamation Plant (NGWRP) in 2001, and in 2002, it began delivering drinking water, about 21,000 m^3 of water per day [75, 76].

19.4 Singapore

This section is an excerpt from Water supply and sanitation in Singapore § Reclaimed water.

Singapore uses NEWater, a brand of reclaimed water, for educational and celebratory reasons. The water is bottled straight from a sophisticated water purification plant. While the majority of Singapore's recycled water goes into the country's high-tech industries, a minor portion is returned to water reservoirs for human use. Reclaimed water is purified to an ultra-pure state and sold under the brand name NEWater.Conventional advanced wastewater treatment facilities, also known as reclamation plants in Singapore, treat wastewater, often known as "used water" in the country. Either microfiltration, reverse osmosis, or UV treatment is applied to the effluent before it is released into the ocean or it is treated further.

The main reason for pursuing water reclaiming was the geopolitical issues that stemmed from Singapore's reliance on water imports from Malaysia.

19.5 South Africa

This section is an excerpt from Water supply and sanitation in South Africa § Wastewater reuse.

Drought is the primary reason why wastewater is being recycled in South Africa [5]. As an example, a direct wastewater reclamation plant (WRP) was built in Beaufort West, South Africa towards the end of 2010 to produce drinking water due to extreme water constraint; the facility has a daily output capacity of 2300 m^3 [77, 78]. Sand filtration, ultrafiltration, two-stage reverse osmosis (RO), and ultraviolet light (UV) disinfection of permeate are all components of the process architecture that is based on the multi-barrier concept.

The town of George implemented an IPR strategy in 2009/2010 in response to water shortages. This strategy involves treating the final effluents from the Outeniqua WWTP with UF and disinfection to a very high standard. The treated effluents are then returned to the Garden Route Dam, the main storage facility, and mixed with the current raw water supplies. This project adds 10,000 m3 to the daily supply, which is enough to meet about a third of the demand for drinking water. The treatment procedures that make up the process configuration include drum screening, ultrafiltration, and chlorine disinfection. George WTW is prepared to apply powdered activated carbon (PAC) as an extra operating barrier if needed [5].

Further following methods may be considered:

- One Water (water management).
- Water conservation.
- Water heat recycling.
- Water recycling shower.
- Water Reuse.

References

1. Tuser C (2022) What is potable reuse? Wastewater Digest. Retrieved 29 Aug 2022
2. Andersson K, Rosemarin A, Lamizana B, Kvarnström E, McConville J, Seidu R, Dickin S, Trimmer C (2016) Sanitation, wastewater management and sustainability: from waste disposal to resource recovery. United Nations Environment Programme and Stockholm Environment Institute, Nairobi and Stockholm. ISBN 978-92-807-3488-1
3. Warsinger DM, Chakraborty S, Tow EW, Plumlee MH, Bellona C, Loutatidou S, Karimi L, Mikelonis AM, Achilli A, Ghassemi A, Padhye LP, Snyder SA, Curcio S, Vecitis CD, Arafat HA, Lienhard JH (2018) A review of polymeric membranes and processes for potable water reuse. Progr Polym Sci 81: 209–237. ISSN 0079-6700. PMC 6011836. PMID 29937599. https://doi.org/10.1016/j.progpolymsci.2018.01.004
4. Takman M, Svahn O, Paul C, Cimbritz M, Blomqvist S, Struckmann Poulsen J, Lund Nielsen J, Davidsson Å (2023) Assessing the potential of a membrane bioreactor and granular activated carbon process for wastewater reuse—a full-scale WWTP operated over one year in Scania, Sweden. Sci Total Environ 895:165185. Bibcode: 2023ScTEn.89565185T. ISSN 0048–9697. PMID 37385512. https://doi.org/10.1016/j.scitotenv.2023.165185
5. Burgess J. Meeker M, Minton J, O'Donohue M (2015) International research agency perspectives on potable water reuse. Environ Sci Water Res Technol 1(5):563–580. https://doi.org/10.1039/C5EW00165J.ISSN2053-1419
6. Jones ER, van Vliet MTH, Qadir M, Bierkens MFP (2021) Country-level and gridded estimates of wastewater production, collection, treatment and reuse. Earth Syst Sci Data 13(2):237–254. Bibcode: 2021ESSD...13..237J. ISSN 1866-3508. https://doi.org/10.5194/essd-13-237-2021
7. US EPA OW (2019) Basic information about water reuse. www.epa.gov. Retrieved 11 May 2022
8. Water Reuse—an overview. ScienceDirect Topics. www.sciencedirect.com. Retrieved 11 May 2022
9. Schiller A (2021) Water reuse. Umweltbundesamt. Retrieved 11 May 2022
10. WHO (2006) WHO guidelines for the safe use of wastewater, excreta and greywater—volume IV: excreta and greywater use in agriculture. World Health Organization (WHO), Geneva, Switzerland
11. WWAP (United Nations World Water Assessment Programme) (2017) The United Nations world water development report 2017. Wastewater: the untapped resource. Paris. ISBN 978-92-3-100201-4. Archived from the original on 08 Apr 2017
12. Direct potable reuse: benefits for public water supplies, agriculture, the environment, and energy conservation (PDF). Retrieved 29 Jul 2016
13. Creel L (2003) Ripple effects: population and coastal regions (PDF) (Report). Population Reference Bureau. Retrieved 29 Jul 2016
14. Guidelines for water reuse (PDF). USEPA. Retrieved 29 Jul 2016

15. European Commission (2016) Joint Research Centre 2014 water reuse in Europe—relevant guidelines, needs for and barriers to innovation. Publications Office. ISBN 9789279443992. https://doi.org/10.2788/29234. Retrieved 29 Jul 2016
16. Tow EW, Hartman AL, Jaworowski AZ, Ines KS, Azadi Aghdam M, Blatchley ER, Achilli A, Gu H, Urper GM, Warsinger DM (2021) Modeling the energy consumption of potable water reuse schemes. Water Res X. 13:100126. Bibcode: 2021WRX....1300126T. ISSN 2589-9147. PMC 8640112. PMID 34901816. S2CID 244483455. https://doi.org/10.1016/j.wroa.2021.100126
17. Moreira da Silva M, Resende FC, Freitas B, Aníbal J, Martins A, Duarte A (2022) Urban wastewater reuse for citrus irrigation in Algarve, Portugal—environmental benefits and carbon fluxes. Sustainability 14(17):10715. Hdl: 10400.1/18203. https://doi.org/10.3390/su141710715
18. Lopes AR, Becerra-Castro C, Vaz-Moreira I, Silva MEF, Nunes OC, Manaia CM (2015) Irrigation with treated wastewater: potential impacts on microbial function and diversity in agricultural soils. In: Wastewater reuse and current challenges. The handbook of environmental chemistry, vol 44. Springer, pp 105–128. HDL: 10400.14/23023. ISBN 978-3-319-23891-3. https://doi.org/10.1007/698_2015_346
19. Bischel HN, Lawrence JE, Halaburka BJ, Plumlee MH, Bawazir AS, King JP, McCray JE, Resh VH, Luthy RG (2013) Renewing urban streams with recycled water for streamflow augmentation: hydrologic, water quality, and ecosystem services management. Environ Eng Sci 30(8):455–479. https://doi.org/10.1089/ees.2012.0201
20. Ritchie R, Mispy O-O (2018) Measuring progress towards the Sustainable Development Goals (SDG 6). SDG-Tracker.org, website
21. National Water Quality Management Strategy (PDF). Retrieved 29 Jul 2016
22. Water recycling and reuse: the environmental benefits. USEPA. Retrieved 29 Jul 2016
23. Water reuse in Europe. Relevant guidelines, needs for and barriers to innovation (PDF). European Union. Retrieved 29 Jul 2016
24. Otoo M, Drechsel P (2018) Resource recovery from waste: business models for energy, nutrient and water reuse in low- and middle-income countries. Routledge—Earthscan, Oxon, UK
25. Garcia-Garcia G, Jagtap S (2021) Enhancement of a spent irrigation water recycling process: a case study in a food business. Appl Sci **11**(21):10355. ISSN 2076-3417. https://doi.org/10.3390/app112110355
26. ISO 16075-1 (2015) Guidelines for treated wastewater use for irrigation projects—Part 1: the basis of a reuse project for irrigation. ISO 21 Mar 2018
27. Ofori S, Puškáčová A, Růžičková I, Wanner J (2021) Treated wastewater reuse for irrigation: pros and cons. Sci Total Environ 760:144026. Bibcode: 2021ScTEn.760044026O. ISSN 0048-9697. PMID 33341618. S2CID 229341652. https://doi.org/10.1016/j.scitotenv.2020.144026
28. Gerrity D, Pecson B, Trussell RS, Trussell RR (2013) Potable reuse treatment trains throughout the world (PDF). J Water Supply: Res Technol Aqua 62(6):321–338. https://doi.org/10.2166/aqua.2013.041. Archived from the original (PDF) on 5 Jul 2016. Retrieved 29 Jul 2016
29. Kean S (2015) Waste not, want not. Distillations 1(4):5. Retrieved 22 Mar 2018
30. Owens B (2015) Pharmaceuticals in the environment: a growing problem. Pharm J. Retrieved 3 Jan 2017
31. Michael-Kordatou I, Michael C, Duan X, He X, Dionysiou DD, Mills MA, Fatta-Kassinos D (2015) Dissolved effluent organic matter: characteristics and potential implications in wastewater treatment and reuse applications. Water Res 77:213–248. Bibcode:2015WatRe..77..213M. PMID 25917290. https://doi.org/10.1016/j.watres.2015.03.011
32. University of Colorado
33. Scientific American Frontiers. Scientific American Frontiers—PBS Programs. Retrieved 12 Mar 2016
34. Astronauts drink recycled urine, and celebrate. Space.com. 20 May 2009
35. Public Utilities Board, Overseas Experiences. Accessed 24 Apr 2007
36. Zhang SX, Babovic V (2012) A real options approach to the design and architecture of water supply systems using innovative water technologies under uncertainty. J Hydroinf 14:13–29. https://doi.org/10.2166/hydro.2011.078

37. Water Scarcity, a driver for water reclamation, reuse and collaboration (PDF). Retrieved 17 Aug 2016
38. Water Reuse—Environment—European Commission. ec.europa.eu. Retrieved 17 Aug 2016
39. Pintilie L, Torres CM, Teodosiu C, Castells F (2016) Urban wastewater reclamation for industrial reuse: an LCA case study. J Clean Prod 139:1–14. Bibcode: 2016JCPro.139....1P. ISSN 0959-6526. https://doi.org/10.1016/j.jclepro.2016.07.209
40. Burgess J, Meeker M, Minton J, O'Donohue M (2015) International research agency perspectives on potable water reuse. Environ Sci Water Res Technol 1(5):563–580. https://doi.org/10.1039/C5EW00165J
41. Wester J, Timpano KR, Çek D, Broad K (2016) The psychology of recycled water: factors predicting disgust and willingness to use. Water Resour Res 52(4):3212–3226. Bibcode:2016WRR....52.3212W. ISSN 1944-7973. https://doi.org/10.1002/2015WR018340
42. Helgeson T (2009) A reconnaissance-level quantitative comparison of reclaimed water, surface water, and groundwater. Water Reuse Research Foundation, Alexandria, VA, p 141. ISBN 978-1-934183-12-0
43. Crook J (2005) Irrigation of parks, playgrounds, and schoolyards: extent and safety. The Water Research Foundation, Alexandria, VA, p 60. ISBN 978-0-9747586-3-3
44. Water Reuse (2012) Water reuse: potential for expanding the nation's water supply through reuse of municipal wastewater. National Research Council. ISBN 978-0-309-25749-7. https://doi.org/10.17226/13303
45. US Environment Protection Agency (2016) Water recycling and reuse: the environmental benefits. Retrieved 22 Aug 2016
46. Wastewater resource recovery can fix water insecurity and cut carbon emissions. European Investment Bank. Retrieved 29 Aug 2022
47. Competing for clean water has led to a crisis. Environment 26 Jan 2010. Archived from the original on 19 Feb 2021. Retrieved 29 Aug 2022
48. Freshwater resources. education.nationalgeographic.org. National Geographic Society. Retrieved 29 Aug 2022
49. Voulvoulis N (2018) Water reuse from a circular economy perspective and potential risks from an unregulated approach. Curr Opin Environ Sci Health 2:32–45. Bibcode: 2018COESH...2...32V. HDL:10044/1/57005. ISSN 2468-5844. https://doi.org/10.1016/j.coesh.2018.01.005
50. Article 16
51. Regulation (EU) 2020/741 of the European Parliament and of the Council of 25 May 2020 on minimum requirements for water reuse
52. Alcalde Sanz L, Gawlik B (2014) Water reuse in Europe—relevant guidelines, needs for and barriers to innovation. Publications Office of the European Union. Retrieved 17 Aug 2016
53. New EU guidelines for agriculture water reuse. www.aquatechtrade.com. Retrieved 29 Aug 2022
54. Usc 1251 seq., 1972, Federal Water Pollution Control Act, Enacted by Congress
55. Lauer WC (1993). Denver's direct potable water reuse demonstration project final report (25 volumes). Denver Water and US EPA Cooperative Agreement CS-806821-01-4
56. Environmental Protection Agency. Retrieved 17 Aug 2016
57. USEPA (2012) 2012 Guidelines for Water Reuse (PDF). Retrieved 5 Jul 2014
58. LeChevallier MW. Overview of water reuse technology: pricing considerations related to reclaimed water. Archived from the original (PPT) on 18 Mar 2009. Retrieved 25 Mar 2009
59. Irvine Ranch Water District. Water reclamation. Archived from the original on 4 Jun 2008. Retrieved 25 Mar 2009
60. The City of San Diego. Water Department. Rules and regulations for recycled water. Archived from the original on 28 May 2009. Retrieved 25 Mar 2009
61. Increasing Safe and Reliable Water Supplies. Water Reuse. 27 Mar 2015. Retrieved 13 May 2022
62. State Corporation Commission—Clerk's Information System. Commonwealth of Virginia. Retrieved 12 May 2022

63. Khouri N, Kalbermatten JM, Bartone CR. Reuse of wastewater in agriculture: a guide for planners (PDF). Retrieved 29 Jul 2016
64. Rodriguez C, Van Buynder P, Lugg R, Blair P, Devine B, Cook A, Weinstein P (2009) Indirect potable reuse: a sustainable water supply alternative. Int J Environ Res Public Health 6(3):1174–1203
65. Burgess J, Meeker M, Minton J, O'Donohue M (2015) International research agency perspectives on potable water reuse. Environ Sci Water Res Technol 1(5):563–580. ISSN 2053-1419. https://doi.org/10.1039/C5EW00165J. Archived (PDF) from the original on 8 Jul 2024 – via Zenodo
66. Arid Israel recycles waste water on grand scale. Archived from the original on 21 May 2014. Retrieved 12 Mar 2016
67. Rinat Z (2012) UN cites Israeli wastewater treatment plant as global model. Haaretz. Retrieved 17 Apr 2013
68. Melanie Lidman (2010) Wastewater wonders. Jerusalem Post. Retrieved 8 Aug 2010
69. Arid Israel recycles waste water on grand scale—Agricultural commodities—Reuters. Archived from the original on 21 May 2014
70. Namibia: Windhoek has been producing drinking water from its wastewater for 50 years. Veolia. Retrieved 29 Aug 2022
71. Recycling sewage into drinking water is no big deal. They've been doing it in Namibia for 50 years. The World from PRX. Retrieved 29 Aug 2022
72. Rodriguez C, Van Buynder P, Lugg R, Blair P, Devine B, Cook A, Weinstein P (2009) Indirect potable reuse: a sustainable water supply alternative. Int J Environ Res Public Health 6(3):1174–1203. PMC 2672392. PMID 19440440. https://doi.org/10.3390/ijerph6031174
73. Michael-Kordatou I, Michael C, Duan X, He X, Dionysiou DD, Mills MA, Fatta-Kassinos D (2015) Dissolved effluent organic matter: characteristics and potential implications in wastewater treatment and reuse applications. Water Res 77:213–248. PMID 25917290. https://doi.org/10.1016/j.watres.2015.03.011
74. Grabow WOK (2009) Water and health - Volume II. EOLSS Publications. ISBN 978-1-84826-183-9
75. Menge J (2010) Treatment of wastewater for re-use in the drinking water system of Windhoek (PDF) (Report)
76. Microbiological risk assessment of the water reclamation plant in Windhoek, Namibia (PDF). Retrieved 29 Jul 2016
77. Risk assessment for South Africa's first direct wastewater reclamation system for drinking water production (PDF). Retrieved 29 Jul 2016
78. Beaufort West water reclamation plant: first direct (toilet-to-tap) water reclamation plant in South Africa (PDF). Retrieved 29 Jul 2016

Conclusions

Biman Gati Gupta

Abstract Water reuse is becoming more popular globally as a means to reduce effluent discharges into receiving waterways, free up scarce potable water resources, and provide a consistent alternate supply of water for uses that do not need high-quality water.

Water reuse is becoming more popular globally as a means to reduce effluent discharges into receiving waterways, free up scarce potable water resources, and provide a consistent alternate supply of water for uses that do not need high-quality water. Agricultural irrigation, non-potable urban and recreational reuse, industrial reuse, collecting and reusing rainfall or runoff, augmenting surface water, recharging groundwater, and potable reuse are among the many water reuse uses across the world. Watering Plants for Agriculture Utilising recycled water for agricultural irrigation is a typical use case. While the use of treated wastewater for agricultural irrigation is on the rise globally, it is most common in water-scarce areas of the Mediterranean and the Middle East. It is common practice to classify water reuse for agricultural irrigation as either limited or uncontrolled. The former makes use of less potable water in some agricultural contexts, and it encompasses irrigating livestock, fibre, seed crops, pastures, commercial nurseries, sod farms, turfgrass, and commercial aquaculture, among other uses.

The usage of water reuse and recycling is still in its early stages in Canada and other parts of world, but the country's experiences may teach us what the future holds. As in other parts of the globe, agricultural and landscape irrigation have been the primary applications of water reuse in Canada. Although there have been reports of wastewater irrigation of golf courses in Ontario, this method is mostly used in British Columbia and the Prairie Provinces. Considerations like as water consumption, nitrogen loading, salinity, and the existence and longevity of pathogens and trace pollutants are essential for effluent irrigation projects. Use of effluent thermal

B. G. Gupta (✉)
Elitte College of Engineering, MAKA University of Technology Sodepur, Kolkata 700113, India
e-mail: bimangupta@yahoo.in

B. G. Gupta (ed.), *Urban Development with Treated Recycled and Reuse of Wastewater: Sustainable Management*, Urban Sustainability,
https://doi.org/10.1007/978-981-96-0857-7_12

energy for snow-making and other non-potable urban reuse applications are also possible. Studies on the reuse of greywater in homes have shown that, while not as polluted as blackwater, there are still quite a few bacteria in greywater. Heavy metals, faecal bacteria, and other pollutants may be present in rainwater collected from roof surfaces. Wastewater treatment systems, when operated and maintained properly, may minimise the amount of these pollutants, and reusing greywater and rainfall can result in significant water savings. Several pilot programmes have been detailed. A wide range of Canadian industries engage in industrial recycling, however to varied degrees. Reusable water quality standards are often process-and application-specific, necessitating the adoption of sophisticated treatment systems to ensure the water meets acceptable chemical and microbiological standards. Further implementation of industrial water reuse and recycling has the potential to significantly reduce Canada's water consumption, since industry is the country's biggest water consumer. Wetland habitat restoration and upstream withdrawal compensation are two potential uses for surface water augmentation. Artificial recharge utilising recycled water is not being used in Canada, despite the country's relatively considerable reliance on groundwater resources. Although it is still done in certain parts of the globe, potable reuse isn't as appealing as non-potable reuse because of the unknown long-term health implications and the existence of new pollutants. Water conservation and reduced effluent emissions into receiving waters are possible outcomes of any of the aforementioned uses of water reuse. The treatment process determines the quality of reclaimed water, and there are many types of reuse that pose different dangers to human and environmental health. Any usage of water reuse must take public health and environmental sustainability into serious consideration.

Bibliography

1. Tom AP, Jayakumar JS, Biju M, Somarajan J, Ibrahim MA (2021) Aquaculture wastewater treatment technologies and their sustainability: a review. Energy Nexus 4(100022)
2. Lata S, Siddharth (2021) Sustainable and eco-friendly approach for controlling industrial wastewater quality imparting succour in water-energy nexus system. Energy Nexus 3(100020)
3. Senfter L, Fritsch M, Berger T, Kofler C, Mayerl M, Pillei M (2021) Kraxner, sludge thickening in a wastewater treatment plant using a modified hydrocyclone carbon Resour. Convers 4:132–141
4. Wang J, Ji Y, Zhang F, Wang D, He X, Wang C (2019) Treatment of coking wastewater using oxic-anoxic-oxic process followed by coagulation and ozonation. Carbon Resour Convers 2:pp 151–156
5. McKinney KR, Gong YY, Lewis TG (2006) Environmental transmission of SARS at Amoy Gardens. J Environ Health 68:26
6. Elsaid K, Olabi V, Sayed ET, Wilberforce T, Abdelkareem MA (2021) Effects of COVID-19 on the environment: an overview on air, water, wastewater, and solid waste. J Environ Manag 292(112694)
7. Article 105689 (8) Alwadani N, Fatehi P (2018) Synthetic and lignin-based surfactants: challenges and opportunities. Carbon Resour Convers 1:126–138

8. Kakavandi B, Ahmadi M (2019) Efficient treatment of saline recalcitrant petrochemical wastewater using heterogeneous UV-assisted sono-Fenton process. Ultrason Sonochem 56:25–36
9. Teodosiu C, Fiore S (2019) Environmental and energy assessment of municipal wastewater treatment plants in Italy and Romania: a comparative study. Water 11:1611
10. Padilla-Rivera A, Güereca LP (2019) A proposal metric for sustainability evaluations of wastewater treatment systems (SEWATS). Ecol Indic 103:22–33
11. Maria FD, Daskal S, Ayalon O (2020) A methodological approach for comparing waste water effluent's regulatory and management frameworks based on sustainability assessment. Ecol Indic 118(106805)
12. Hijazi O, Tappen S, Effenberger M (2019) Environmental impacts concerning flexible power generation in a biogas production. Carbon Resour Convers 2:117–125
13. Sfez S, De Meester S, Vlaeminck SE, Dewulf J (2019) Improving the resource footprint evaluation of products recovered from wastewater: a discussion on appropriate allocation in the context of circular economy. Resour Conserv Recycl 148:132–144
14. U. Nations, The sustainable development goals report 2019 United Nations (2019)
15. Roychand R, Kumar Pramanik B, Zhang G, Setunge S (2020) Recycling steel slag from municipal wastewater treatment plants into concrete applications–a step towards circular economy. Resour Conserv Recycl 152(104533)
16. Wilberforce T, Olabi AG, Sayed ET, Elsaid K, Abdelkareem MA (2021) Progress in carbon capture technologies. Sci Total Environ 761(143203)
17. Solarin SA (2020) An environmental impact assessment of fossil fuel subsidies in emerging and developing economies. Environ Impact Assess Rev 85(106443)
18. Olabi V, Wilberforce T, Elsaid K, Sayed ET, Abdelkareem MA (2022) Impact of COVID-19 on the renewable energy sector and mitigation strategies. Chem Eng Technol 45:558–571
19. Foteinis AG, Borthwick Z, Frontistis D, Mantzavinos E (2018) Chatzisymeon Environmental sustainability of light-driven processes for wastewater treatment applications. J Clean Prod 182:8–15
20. Olabi AG, Obaideen K, Elsaid K, Wilberforce T, Sayed ET, Maghrabie HM, Abdelkareem MA (2022) Assessment of the pre-combustion carbon capture contribution into sustainable development goals SDGs using novel indicators. Renew Sustain Energy Rev 153(111710)
21. Obaideen K, AlMallahi MN, Alami AH, Ramadan M, Abdelkareem MA, Shehata N, Olabi AG (2021) On the contribution of solar energy to sustainable developments goals: case study on Mohammed bin Rashid Al Maktoum Solar Park. Int J Thermo Fluids 12(100123)
22. Ghisellini P, Cialani C, Ulgiati S (2016) A review on circular economy: the expected transition to a balanced interplay of environmental and economic systems. J Clean Prod 114:11–32
23. Shehata N, Mohamed OA, Sayed ET, Abdelkareem MA, Olabi AG (2022) Geopolymer concrete as green building materials: recent applications, sustainable development and circular economy potentials. Sci Total Environ 836
24. Neczaj E, Grosser A (2018) Circular economy in wastewater treatment plant-challenges and barriers multidisciplinary digital publishing institute proceedings, p 614
25. Alcamo J (2019) Water quality and its interlinkages with the sustainable development goals. Curr Opin Environ Sustain 36:126–140
26. Tortajada C (2020) Contributions of recycled wastewater to clean water and sanitation sustainable development goals. NPJ Clean Water 3:22
27. Delanka-Pedige HMK, Munasinghe-Arachchige SP, Abeysiriwardana-Arachchige ISA, Nirmalakhandan N (2021) Wastewater infrastructure for sustainable cities: assessment based on UN sustainable development goals (SDGs). Int J Sustain Dev World Ecol 28:203–209
28. Delanka-Pedige HMK, Munasinghe-Arachchige SP, Abeysiriwardana-Arachchige ISA, Nirmalakhandan N (2021) Evaluating wastewater treatment infrastructure systems based on UN sustainable development goals and targets. J Clean Prod 298(126795)
29. Palme U, Lundin M, Tillman AM, Molander S (2005) Sustainable development indicators for wastewater systems–researchers and indicator users in a co-operative case study. Resour Conserv Recycl 43:293–311

30. Sayed ET, Abdelkareem MA, Obaideen K, Elsaid K, Wilberforce T, Maghrabie HM, Olabi AG (2021) Progress in plant-based bioelectrochemical systems and their connection with sustainable development goals. Carbon Resour Convers 4:169–183
31. Tan C, Zhi Q (2016) The energy-water nexus: a literature review of the dependence of energy on water. Energy Procedia 88:277–284
32. Li Z, Xu X, Sheng X, Lin P, Tang P, Pan L, Kaneti YV, Yang T, Yamauchi Y (2021) Solar-powered sustainable water production: state-of-the-art technologies for sunlight-energy-water nexus ACS Nano 15:12535–12566
33. Chini CM, Stillwell AS (2018) The state of US urban water: data and the energy-water nexus. Water Resour Res 54:1796–1811
34. DeNooyer TA, Peschel JM, Zhang Z, Stillwell AS (2016) Integrating water resources and power generation: the energy-water nexus in Illinois. Appl Energy 162:363–371
35. Zhang X, Vesselinov VV (2016) Energy-water nexus: balancing the tradeoffs between two-level decision makers. Appl Energy 183:77–87
36. Stillwell AS, King CW, Webber ME, Duncan IJ, Hardberger A (2011) The energy-water nexus in Texas. Ecol Soc 16
37. D'Odorico P, Davis KF, Rosa L, Carr JA, Chiarelli D, Dell'Angelo J, Gephart J, MacDonald GK, Seekell DA, Suweis S (2018) The global food-energy-water nexus. Rev Geophys 56:456–531
38. Chen S, Chen B (2016) Urban energy-water nexus: a network perspective. Appl Energy 184:905–914
39. Bazilian M, Rogner H, Howells M, Hermann S, Arent D, Gielen D, Steduto P, Mueller A, Komor P, Tol RS (2011) Considering the energy, water and food nexus: towards an integrated modelling approach. Energy Policy 39:7896–7906
40. Hussey K, Pittock J (2012) The energy-water nexus: managing the links between energy and water for a sustainable future. Ecol Soc, p 17
41. Zaher A, Shehata N (2021) Recent advances and challenges in management of urea wastewater: a mini review. In: Proceedings of the IOP conference series: materials science and engineering, IOP Publishing, Article 012021
42. Egirani D, Poyi N, Shehata N (2020) Preparation and characterization of powdered and granular activated carbon from Palmae biomass for cadmium removal. Int J Environ Sci Technol 17:2443–2454
43. Egirani D, Latif MT, Wessey N, Poyi NR, Shehata N (2021) Preparation and characterization of powdered and granular activated carbon from Palmae biomass for mercury removal. Appl Water Sci 11:1–11
44. Moustafa D, Mahmoud R, Abd El-Salam HM, Shehata N (2021) Utilization of residual zinc-iron-layered double hydroxide after methyl orange management as a new sorbent for wastewater treatment. Appl Nanosci 11:709–723
45. Collivignarelli MC, Abbà A, Miino MC, Damiani S (2019) Treatments for color removal from wastewater: state of the art. J Environ Manag 236:727–745
46. Mathuram M, Meera R, Vijayaraghavan G (2018) Application of locally sourced plants as natural coagulants for dye removal from wastewater: a review. J Mater Environ Sci 2508:2058–2070
47. Abdullah N, Yusof N, Lau W, Jaafar J, Ismail A (2019) Recent trends of heavy metal removal from water/wastewater by membrane technologies. J Ind Eng Chem 76:17–38
48. Pan Z, Song C, Li L, Wang H, Pan Y, Wang C, Li J, Wang T, Feng X (2019) Membrane technology coupled with electrochemical advanced oxidation processes for organic wastewater treatment: recent advances and future prospects. Chem Eng J 376(120909)
49. Arif AUA, Sorour MT, Aly SA (2020) Cost analysis of activated sludge and membrane bioreactor WWTPs using capdetworks simulation program: case study of Tikrit WWTP (middle Iraq Alex. Eng J 59:4659–4667
50. Pandey A, Singh P, Iyengar L (2007) Bacterial decolorization and degradation of azo dyes Int. Biodeterior. Biodegrad 59:73–84

51. Popli S, Patel UD (2015) Destruction of azo dyes by anaerobic-aerobic sequential biological treatment: a review. Int J Environ Sci Technol 12:405–420
52. Jafari N, Soudi MR, Kasra-Kermanshahi R (2014) Biodegradation perspectives of azo dyes by yeasts. Microbiology 83:484–497
53. Solís M, Solís A, Pérez HI, Manjarrez N, Flores M (2012) Microbial decolouration of azo dyes: a review. Process Biochem 47:1723–1748
54. Katheresan V, Kansedo J, Lau SY (2018) Efficiency of various recent wastewater dye removal methods: a review. J Environ Chem Eng 6:4676–4697
55. Lee KE, Morad N, Teng TT, Poh BT (2012) Development, characterization and the application of hybrid materials in coagulation/flocculation of wastewater: a review. Chem Eng J 203:370–386
56. Cai Q, Lee B, Ong S, Hu J (2020) Fluidized-bed Fenton technologies for recalcitrant industrial wastewater treatment-recent advances, challenges and perspective. Water Res, 116692
57. Tafti AD, Mirzaii SMS, Andalibi MR, Vossoughi M (2015) Optimized coupling of an intermittent DC electric field with a membrane bioreactor for enhanced effluent quality and hindered membrane fouling Sep. Purif Technol 152:7–13
58. Hakimhashemi M, Gebreyohannes AY, Saveyn H, Van der Meeren P, Verliefde A (2012) Combined effects of operational parameters on electro-ultrafiltration process characteristics. J Membr Sci 403:227–235
59. Sarkar B, De B (2011) Prediction of permeate flux for turbulent flow in cross flow electric field assisted ultrafiltration. J Membr Sci 369:77–87
60. Tsai YT, Weng YH, Lin AYC, Li KC (2011) Electro-microfiltration treatment of water containing natural organic matter and inorganic particles. Desalination 267:133–138
61. Xu L, Du LS, Wang C, Xu W (2012) Nanofiltration coupled with electrolytic oxidation in treating simulated dye wastewater. J Membr Sci 409:329–334
62. Dudchenko AV, Rolf J, Russell K, Duan W, Jassby D (2014) Organic fouling inhibition on electrically conducting carbon nanotube-polyvinyl alcohol composite ultrafiltration membranes. J Membr Sci 468:1–10
63. Liu S, Wang Y, Zhou X, Han W, Li J, Sun X, Shen J, Wang L (2017) Improved degradation of the aqueous flutriafol using a nanostructure macroporous PbO_2 as reactive electrochemical membrane. Electrochim Acta 253:357–367
64. Ćemanović A, Arikan OA, Kizilet A, Çınar Ö (2018) Treatment of industrial wastewaters with anaerobic membrane bioreactors. In: Proceedings of the 4th international conference on engineering and natural sciences at Kiev, Ukraine
65. Chang IS, Clech PL, Jefferson B, Judd S (2002) Membrane fouling in membrane bioreactors for wastewater treatment. J Environ Eng 128:1018–1029
66. Karabelas AJ, Plakas KV, Sarasidis VC (2018) How far are we from large-scale PMR applications? current trends and future developments on (Bio-) membranes, Elsevier, pp 233–295
67. Moslehyani A, Hubadillah SK, Othman MHD, Ismail AF, Matsuura T (2018) PMRs in photodegradation of organic contaminants: water and wastewater treatment. Current Trends Future Developments On (Bio-) Membranes, Elsevier, pp 189–208
68. Zheng X, Shen ZP, Shi L, Cheng R, Yuan DH (2017) Photocatalytic membrane reactors (PMRs) in water treatment: configurations and influencing factors. Catalysts 7:224
69. Molinari R, Palmisano L, Drioli E, Schiavello M (2002) Studies on various reactor configurations for coupling photocatalysis and membrane processes in water purification. J Membr Sci 206:399–415
70. Sangave PC, Pandit AB (2006) Ultrasound and enzyme assisted biodegradation of distillery wastewater. J Environ Manag 80:36–46
71. Dehghani MH, Zarei A, Yousefi M (2019) Efficiency of ultrasound for degradation of an anionic surfactant from water: surfactant determination using methylene blue active substances method MethodsX 6:805–814
72. Hilares RT, Atoche-Garay DT, Pagaza DAP, Ahmed MA, Andrade GJC, Santos JC (2021) Promising physicochemical technologies for poultry slaughterhouse wastewater treatment: a critical review. J Environ Chem Eng 9(105174). https://doi.org/10.1016/j.jece.2021.105174

73. Suslick KS, Mdleleni MM, Ries JT (1997) Chemistry induced by hydrodynamic cavitation. J Am Chem Soc 119:9303–9304
74. Kumar PS, Pandit A (1999) Modeling hydrodynamic cavitation. Chem Eng Technol 22:1017–1027 Industrial Chemistry-Plant Equipment-Process Engineering-Biotechnology
75. Gogate P, Thanekar P, Oke A (2020) Strategies to improve biological oxidation of real wastewater using cavitation based pre-treatment approaches. Ultrason Sonochem 64(105016)
76. Padoley K, Saharan VK, Mudliar S, Pandey R, Pandit AB (2012) Cavitationally induced biodegradability enhancement of a distillery wastewater. J Hazard Mater 219:69–74
77. Bis M, Montusiewicz A, Ozonek J, Pasieczna-Patkowska S (2015) Application of hydrodynamic cavitation to improve the biodegradability of mature landfill leachate. Ultrason Sonochem 26:378–387
78. Arora S, Saraswat S (2021) Vermifiltration as a natural, sustainable and green technology for environmental remediation: a new paradigm for wastewater treatment process. Curr Res Green Sustain Chem 4(100061)
79. Arora S, Kazmi A (2015) The effect of seasonal temperature on pathogen removal efficacy of vermifilter for wastewater treatment. Water Res 74:88–99
80. Lourenço N, Nunes L (2017) Is filter packing important in a small-scale vermifiltration process of urban wastewater?. Int J Environ Sci Technol 14:2411–2422
81. Rajpal A, Arora S, Bhatia A, Kumar T, Bhargava R, Chopra A, Kazmi A (2014) Co-treatment of organic fraction of municipal solid waste (OFMSW) and sewage by vermireactor. Ecol Eng 73:154–161
82. Tomar P, Suthar S (2011) Urban wastewater treatment using vermi-biofiltration system. Desalination 282:95–103
83. Samal K, Dash RR, Bhunia R (2017) Performance assessment of a Canna indica assisted vermifilter for synthetic dairy wastewater treatment. Process Saf Environ Prot 111:363–374
84. Singh R, D'Alessio M, Meneses Y, Bartelt-Hunt SL, Woodbury B, Ray C (2021) Development and performance assessment of an integrated vermifiltration based treatment system for the treatment of feedlot runoff. J Clean Prod 278(123355)
85. Kumar C, Ghosh AK (2019) Fabrication of a vermifiltration unit for wastewater recycling and performance of vermifiltered water (vermiaqua) on onion (Allium cepa). Int J Recycl Org Waste Agric 8:405–415
86. Balkema A, Weijers S, Lambert F (1998) On methodologies for comparison of wastewater treatment systems with respect to sustainability. In: Proceedings of the conference options for closed water systems, pp 11–13
87. Ozsahin DU, Uzun B, Sani M, Ozsahin I (2018) Evaluating X-ray based medical imaging devices with fuzzy preference ranking organization method for enrichment evaluations. Int J Adv Comput Sci Appl 9:7–10
88. Yahya MN, Gökçekuş H, Ozsahin DU (2020) Comparative analysis of wastewater treatment technologies. J Kejuruter 32:221–230
89. Bagheri A, Hjorth P (2007) A framework for process indicators to monitor for sustainable development: practice to an urban water system. Environ Dev Sustain 9:143–161
90. Goffi AS, Trojan F, de Lima JD, Lizot M, Thesari SS (2018) Economic feasibility for selecting wastewater treatment systems. Water Sci Technol 78:2518–2531
91. Kalbar PP, Karmakar S, Asolekar SR (2012) Selection of an appropriate wastewater treatment technology: a scenario-based multiple-attribute decision-making approach. J Environ Manag 113:158–169
92. Zeng G, Jiang R, Huang G, Xu M, Li J (2007) Optimization of wastewater treatment alternative selection by hierarchy grey relational analysis. J Environ Manag 82:250–259
93. Karimi A, Mehrdadi N, Hashemian S, Bidhendi GN, Moghaddam RT (2011) Selection of wastewater treatment process based on the analytical hierarchy process and fuzzy analytical hierarchy process methods. Int J Environ Sci Technol 8:267–280
94. Zhao Y, Qin Y, Chen B, Zhao X, Li Y, Yin X, Chen G (2009) GIS-based optimization for the locations of sewage treatment plants and sewage outfalls-a case study of Nansha District in Guangzhou City, China Commun. Nonlinear Sci Numer Simul 14:1746–1757

95. Gbanie SP, Tengbe PB, Momoh JS, Medo J, Kabba VTS (2013) Modelling landfill location using geographic information systems (GIS) and multi-criteria decision analysis (MCDA): case study Bo. Southern Sierra Leone Appl Geogr 36:3–12
96. Ren J, Liang H, Chan FT (2017) Urban sewage sludge, sustainability, and transition for Eco-City: multi-criteria sustainability assessment of technologies based on best-worst method. Technol Forecast Soc Chang 116:29–39
97. Lizot M, Goffi AS, Thesari SS, Trojan F, Afonso PS, Ferreira PF (2020) Multi-criteria methodology for selection of wastewater treatment systems with economic, social, technical and environmental aspects. Environ Dev Sustain, pp 1–25
98. Sayed ET, Nakagawa N (2018) Critical issues in the performance of yeast based microbial fuel cell. J Chem Technol Biotechnol 93:1588–1594
99. Sophonsiri C, Morgenroth E (2004) Chemical composition associated with different particle size fractions in municipal, industrial, and agricultural wastewaters. Chemosphere 55:691–703
100. Gormley M, Aspray TJ, Kelly DA (2020) COVID-19: mitigating transmission via wastewater plumbing systems. Lancet Glob Health 8:e643
101. Plakas K, Georgiadis A, Karabelas A (2016) Sustainability assessment of tertiary wastewater treatment technologies: a multi-criteria analysis. Water Sci Technol 73:1532–1540
102. Hussain N, Abdelkareem MA, Alawadhi H, Alaswad A, Sayed ET (2021) Two dimensional Cu based nanocomposite materials for direct urea fuel cell. Int J Hydrog Energy 46:6051–6060
103. Alami AH, Abdelkareem MA, Faraj M, Aokal K, Al Safarini N (2020) Titanium dioxide-coated nickel foam photoelectrodes for direct urea fuel cell applications. Energy 208(118253)
104. Vergine P, Salerno C, Libutti A, Beneduce L, Gatta G, Berardi G, Pollice A (2017) Closing the water cycle in the agro-industrial sector by reusing treated wastewater for irrigation. J Clean Prod 164:587–596
105. Cornejo PK, Zhang Q, Mihelcic JR (2016) How does scale of implementation impact the environmental sustainability of wastewater treatment integrated with resource recovery?. Environ Sci Technol 50:6680–6689
106. Gu J, Yang Q, Liu Y (2018) Mainstream anammox in a novel A-2B process for energy-efficient municipal wastewater treatment with minimized sludge production. Water Res 138:1–6
107. De Francisci D, Su Y, Iital A, Angelidaki I (2018) Evaluation of microalgae production coupled with wastewater treatment. Environ Technol 39:581–592
108. Greenberg MR (2013) The environmental impact statement after two generations: managing environmental power Routledge
109. Sundarraj DB, Ruthaevel PMK, Kumaraswamy K (2010) An investigation of groundwater quality and its suitability to irrigated agriculture in Coimbatore district, Tamilnadu, India, A GIS approach. Int J Environ Sci 1(2):176–190
110. Chatterjee D, Lowell K, Rivoire B, McNeil MR, Brennan PJ (1992) Lipoarabinomannan of Mycobacterium tuberculosis. Capping with mannosyl residues in some strains. J Biol Chem 267(9):6234–39
111. Jarvie HP, Whitton BA, Neal C (1998) Nitrogen and phosphorous in east coast British rivers: specification. Sour Bio Sig Sci Total Env 210–211:79–109
112. Kannan V (1993) Tharamani, Assessment of industrial groundwater pollution potential from the correlation of parametric ratio-Dye industry. Ind J Environ Protect 13(5):346–348
113. Sundarraj DB, Ruthaevel PMK, Kumaraswamy K (2010) An investigation of groundwater quality and its suitability to irrigated agriculture in Coimbatore district, Tamilnadu, India, A GIS approach. Int J Environ Sci 1(2):176–190
114. Shakhashiri P (2008) Acetic acid and acetic anhydride. J General Chem, pp 2–3. http://www.scifun.org
115. Joshi VJ, Santani DD (2012) Physio-chemical characterization and heavy metal concentration in effluent of the textile industry. Uni J Env Res Technol 2(2):93–96
116. Dick, JJ, Soulsby C, Birkel C et al (2016) Continuous dissolved oxygen measurements and modelling metabolism in Peatland streams. PLOS One. https://doi.org/10.1371/journal.pone.0161363

117. Prabha S, Kumar M, Kumar A, Das P, Ramanathan AL (2013) Impact assessment of textile effluent on groundwater quality in the vicinity of Tirupur industrial area, southern India. Environ Earth Sci. https://doi.org/10.1007/s12665-013-2361-8
118. Wang Q, Luan Z, Wei N, Li J, Liu C (1997) The color removal of dye wastewater treatment and reuse. In: Proceeding of membrane technologies for wastewater by advance oxidation processes—a review. Global Nest Int J 6
119. Pérez-Estrada L, García-Montaño J, Oller I, Maldonado MI, Torrades F, Peral J (2008) Pilot plant scale reactive dyes degradation by solar photo-Fenton and biological processes. J Photochem Photobiol, A 195(2):205–214

The manufacturer's authorised representative in the EU is Springer
Nature Customer Service Centre GmbH, Europaplatz 3, 69115 Heidelberg,
Germany. If you have any concerns regarding our products, please
contact ProductSafety@springernature.com

Printed and bound by CPI Group (UK) Ltd, Croydon, CR0 4YY
01/10/2026
02225876-0012